The Anthropology of Community-Based Whaling in Greenland

The Anthropology of Community-Based Whaling in Greenland

A Collection of Papers Submitted to the International Whaling Commission

Edited by
Marc G. Stevenson, Andrew Madsen, and
Elaine L. Maloney

Studies in Whaling No. 4
Occasional Publication No. 42
© 1997 Canadian Circumpolar Institute
University of Alberta

Canadian Cataloguing in Publication Data

Main entry under title:

Community-Based Whaling in Greenland

(Occasional publication series, ISSN 0068-0303; no. 42)
ISBN 1-896445-05-5

1. Whaling — Greenland. 2. Inuit — Greenland — Fishing. 3. Inuit — Greenland — Social Life and Customs I. Stevenson, Marc. II. Madsen, Andrew. III. Maloney, Elaine L. IV. International Whaling Commission. V. Canadian Circumpolar Institute. VI. Series: Occasional publication series (Canadian Circumpolar Institute); no. 42.

E99.E7C69 1997   306'.089'97120982   C97-910513-7

Studies in Whaling No. 4
© 1997 Canadian Circumpolar Institute
Cover design by Art Design Printing, Inc.
Printed by Art Design Printing, Inc.

(This book was made possible with funds from the North Atlantic Marine Mammal Commission - NAMMCO Fund, and the Japan Small-Type Whaling Association)

Table of Contents

Editorial Notes XV

INTRODUCTION 1
    Bibliography 15

Chapter 1
ABORIGINAL SUBSISTENCE WHALING:
A CONTEMPORARY OUTLOOK 17
(Moses Olsen, 1988)

    Nutrition 20
    The Need for Management of Whale 'Stocks' 21
    The Need for International Advice on Management 21
    Using Advice in the Formation of Management Decisions 23
    Positive Thinking 25

Chapter 2
SUBSISTENCE WHALING IN GREENLAND 27
(R. Petersen, E. Lemke, and F.O. Kapel, 1981)

    Introduction 29
    General Information on Greenlandic Whaling 30
        Occurrence and Distribution of Subsistence Whaling
            by Indigenous Peoples 30
        Areas in which Subsistence Whaling Occurs 30
        Species Taken and 'Stocks' from which Takes Occur 30
        Methods and Equipment Used (Past And Present) 31
        Catch: Historic and Recent 31

Table of Contents

Commercial Catches from 'Stocks' Where
    Subsistence Whaling Occurs 31
Products and Processing of Subsistence Catches by
    Indigenous Peoples 32
        Products: Processing And Use 32
Nature and Level of Distribution, Barter, and Trade 34
Cultural and Nutritional Attributes of Subsistence
    Whaling by Indigenous Peoples 35
Management Principles Potentially Applicable to
    Subsistence Whaling by Indigenous Peoples 36
Management and Regulation of Whaling 37
        National Regulations 37
        Revised Management Procedures 38
Approaches to Setting Catch Limits for 'Stocks' Where
    Subsistence Whaling Occurs 38
        Scientific Committee Advice 38
        Assessment of Cultural and Nutritional Requirements
            of Indigenous Peoples 39
        Reporting and Research Requirements for
            Subsistence Catches 40
        Information on Composition of Catches and Hunting
            Effort 41
        Information on Utilization of Products and on
            Socioeconomic Values of Hunting 41
Subsistence Whaling in Greenland 42
        General Considerations of Research and Monitoring
            Requirements.. 42
        Independent Observation of Subsistence Catches 43
        Possible Impact on Traditional Hunting 44
        Logistic Arrangements 44
        Inclusion of Subsistence Whaling in an Approved
            IWC Observer Scheme 45
        Examination of Killing Techniques 46
Bibliography 47

Appendix I: Nutritional Needs Relating to Aboriginal
Subsistence Whaling Among the Inuit in Greenland
(by Peder Helms) 49
    The Value of Whale Meat in the Greenland Diet 49
    The Nutritional Value of Whale Meat 50
    List of References 53

Chapter 3
THE GREENLAND ABORIGINAL WHALE HUNT   55
(P. Helms, O. Hertz, and F.O. Kapel, 1984)

Introduction 58

Subsistence Needs 59
    Human Population: Occupational Regions and the
        Distribution of Whaling 59
    Number of Communities Engaged in Whaling and
        Dependence on Whaling 64

The Importance of Whaling 65

Employment Opportunities/Alternatives 66

Details of the Hunt: Historical Review, by Species 67
    Seal Hunting 69
    Walrus and Polar Bear Hunting 69
    Hunting of Small Cetaceans 69
    Caribou Hunting 70
    Muskox Hunting 70
    Fox Hunting 72
    Bird Hunting 72
    Fishing 72
    Hunting of Large Whales 73
    Organization and Seasonal Variation in the Hunt of
        Large Whales 75

Review of the 1978-82 and 1983 Seasons 76

Whale Hunting Methods and Efficiency 79

Cultural Needs 79
    Participation in Preparations for the Whale Hunt,
        the Hunt Itself, and Processing the Products of the
        Hunt 79

   Distribution and Sharing of Whale Meat and
    Products, Food Preferences, and Degree of
    Utilization 80
   Ceremonies, Feasts, and Folklore 80
   Social Integrative Functions of the Whale Hunt and
    Risk to Community Identity from Imposed
    Restrictions on Aboriginal Whaling 81
   Other Uses of Whales 82
  Nutritional Needs 82
   Role of Whale Products as Food in the Community
    and the Importance of Whale Products in the
    Traditional Diet 82
   Alternative Food Resources to Whale Meat and
    Availability and Acceptability of Other Food
    Sources 85
  References 89

Chapter 4
COMMUNAL ASPECTS OF PREPARING FOR WHALING, THE HUNT ITSELF, AND THE ENSUING PRODUCTS   93
(Robert Petersen, 1987)

  Introduction 95
  Distribution, Sharing, and the Role of Money 96
  Social Status, Prestige, and Community 97
  Conclusion 99

Chapter 5
SCORESBYSUND: A HUNTING COMMUNITY IN EAST GREENLAND   101
(Fin B. Larsen, 1987)

  Introduction 103
  The Founding and Historical Development of
   Scoresbysund 105
  The Economic Situation of the Hunting Occupation 107
  Traditional Elements in the Modern Hunting Culture 109

       The Dog Sled 109
       Hunt Sharing Rules... 110
       Food Preferences 111

The Concept of Time and Attitude Towards Work 111
       The Altruistic Way of Dealing with Property 112

Exploitation of Living Resources 112
       Marine Mammals... 114
       Land Mammals 116
       Birds.. 117
       Fish and Sharks... 118

The Use of the Products of the Hunt 118
       Skin 118
       Edible Tissue 120
       Tusks, Bones, etc. 121

Local Production of Food, 1985... 121
       The Importance of Minke Whale Hunting in the
           Economy.. 123

Conclusion 124

Bibliography 125

Chapter 6
GREENLAND SUBSISTENCE HUNTING   127
(Janne Jervin [project coordinator],
with contributions by Jens Dahl, Peder Helms,
and Robert Petersen, 1989)

   Introduction 129
       What Actually is Traditional Whaling? 130

   General Facts About Greenland 133
       Introduction 133
       Geography... 133
       Climate.. 134
       Vegetation... 135
       Ice.. 135
       Population 136
       Language... 136
       Trade and Industry 137
       Hunting Licenses 138

Table of Contents

History 139
    The First Eskimos 139
    Thule and Inussuk 141
    History of Aboriginal Settlement (continued) 143
    The Norsemen and the Inughuit 143
    Colonial era 143
    Integration into Denmark 144
    Greater Self-Determination and Home Rule 145
    Landsting (Greenland's Parliament) and Landsstyre
        (Greenland's Government) 145
    Representatives of the Danish State 146
    Foreign Relations 146
    Greenland and the EEC 147
    Greenland and Nordic Cooperation 147
    The Municipalities 147

Sources of Background Knowledge Concerning the
    Nature of Subsistence in Greenland 148
    Basis of Livelihood: Animals Hunted 148
    Extracts of Reports and Publications Concerning
        Greenland 152
    The *Sonja* 153
    Reports Concerning Greenland (Beretninger
        vedrørende Grønland), 1955: An Examination of
        Greenlandic Foodstuff and Eating Habits, by the
        State Health Investigation Board and Vitamin
        Laboratory… 165
    An interview with the Nutritional and Dietary Expert,
        Dr. Peder Helms, M.D., 'Atuisoq'
        4 December 1988… 166

Hunting and Subsistence in Greenland in Light of
    Socioeconomic Relations (by Jens Dahl) 168
    Introduction 168
    The Structural Importance of Subsistence Activities
        in Greenland 172
    The Traditional Mode of Production 174
    Hunting and Fishing and Subsidiary Activities 176
    Conclusion 178

Traditional and Present Distribution Channels in
   Subsistence Hunting in Greenland (by Robert Petersen) 179
      The Homogeneous Community 179
      Solidarity and Balanced Exchange.................................. 180
      Development of the Greenlandic Community in the
         20th Century... 181
      Exchange and Distribution of Goods Today................... 182
      The Role of Money and the Commercialization
         of Hunting 183
      Implements Used for Hunting 183
      Some Aspects of Distribution 184
      Subsistence or Commercial Hunting?............................. 184
      Symbiotic Relationship 185
      Free Distribution Today 186
      Conclusion 187
   References 188

Chapter 7
INUIT AND WHALES AT SARFAQ (GREENLAND)   189
(Svend E. Larsen and Klaus G. Hansen, 1990)

Sarfaq.. 192
   The Socioeconomic Structure of the Settlement............... 193

A Collective Whale Hunt 201

Whaling by Cutter 208
   Flensing a Whale 208

Distributing the Catch after a Collective Whale Hunt............ 209

Distribution of a Whale Caught During a Cutter Hunt............ 212

Households Involved in Whaling Activities 214

Utilization, Distribution, and Storage of Whale Products 215

Conclusion 218

Bibliography 219

Table of Contents

Chapter 8
CUTTER HUNTING OF MINKE WHALE
IN QAQORTOQ (GREENLAND)   221
(Erling Josefsen, 1990)

    Introduction 223
    Whalers and Their Crew 224
    The Whaling Season 226
    The Hunt 227
          The Following Morning 229
    The Flensing 230
    Distribution of Whale Products 231
    Sale of Whale Products 232
    Distribution and Consumption of Whale Products 234
    Bibliography 235
    Appendix I: Historical Background 236
          Greenlandic Natural Whaling of Bowhead and
              Humpback 236
          Whaling in the 20th Century 237
          The Tovqussaq Whaling Station 238

Chapter 9
GREENLAND INUIT WHALING IN
QEQERTARSUAQ KOMMUNE   239
(Richard A. Caulfield, 1991)

    Nutritional and Sociocultural Significance of Whales in
        Qeqertarsuaq Kommune 241
          Overview 241
          Nutritional Significance of Minke and Fin Whale
              Products 242
          Sharing of Wild Foods in Households and
              Communities 245
          Wild Foods in Household and Community
              Celebrations 247
          Whaling and Greenlandic Culture Identity 249

Analysis of Contemporary Whaling Issues in
   Qeqertarsuaq Kommune 252
      Overview.. 252
      Profit Maximization and Commodification in Whaling .. 252
      Capital Intensification and New Technology in
         Whaling 254
      Impact of External Regulatory Regimes on Whaling
         in Qeqertarsuaq Kommune 254
      Conclusion 257
Bibliography 258

Chapter 10
WHALING AND SUSTAINABILITY IN GREENLAND 261
(Richard A. Caulfield, 1994)

Introduction 263
Whaling Within Socially-Defined Groups:
   Greenland's Inuit Society 264
Whaling Within Territorial Limits: Coastal Catches
   for Local Consumption 266
Social Reproducibility: Greenlandic Whaling
   Through the Generations 267
Valuing Greenlandic Whaling Multi-Dimensionally 269
Biological Sustainability and Greenlandic Whaling................. 271
Whaling and Indigenous Rights 274
Thinking Globally, Acting Locally: Sustainable
   Whaling in Greenland 275
References 276

Editorial Notes

Editing a series of papers written in English mostly by authors for whom Danish and *Kalaallisut* (the Inuit language in Greenland) are the primary languages, and English a second or third language, presents certain challenges. These are compounded by the fact that these papers were written over a period of a dozen years with different intentions and motivations, all of which were responses to IWC requests, decisions, and actions.

Where possible, an attempt had been made to present the papers in this volume in a consistent format. In regards to Inuit spelling, some authors spell words such as whale skin (e.g., *muktuk, mattak*) or place names (e.g., Ammassalik, Angmagssalik) differently. We have chosen not to standardize such spellings in order to respect the author(s) preference, and because these differences may reflect regional dialectal variation. Neither are bibliographies and/or lists of references at the back of each contribution presented in a consistent manner owing to the stylistic differences of the authors, although an effort was made to standardize and include as much information as possible in each citation.

Inuit words not yet captured into the English language (e.g., *muktuk, umiat*) are italicized, whereas Inuit place names (e.g., Sarfaq, Qeqertarsuaq, etc.) are not. English terms such as 'stock', 'harvest', 'management', etc. are commonly used in the context of 'wildlife management' as practised by various nation states and the IWC. However, increasingly Greenlanders are growing dissatisfied with the use of such concepts to describe their relationship with whales and other marine mammals upon which they depend (Egede 1995). Although many of the authors in this volume employ such terms, their use should be viewed in both historical context and in cross-cultural context, where they have different meanings and values to different people.

Numerous tables, figures, and photographs accompanying the various manuscripts submitted to the IWC have not been included in this volume owing either to lack of availability or for the sake of economy and consistency. While regrettable — they contributed significantly to the richness and flavour of the originals — only those that were considered integral to the discussion have been included in this volume. Several manuscripts have also been extensively edited and shortened. For complete versions of the originals, the reader is urged to contact the International Whaling Commission or the Greenland Home Rule authorities.

Reference

Egede, I. 1995. Paper presented at *Circumpolar Aboriginal People and Co-management Practice*. Inuvik, NWT, Canada, 20-23 November 1995.

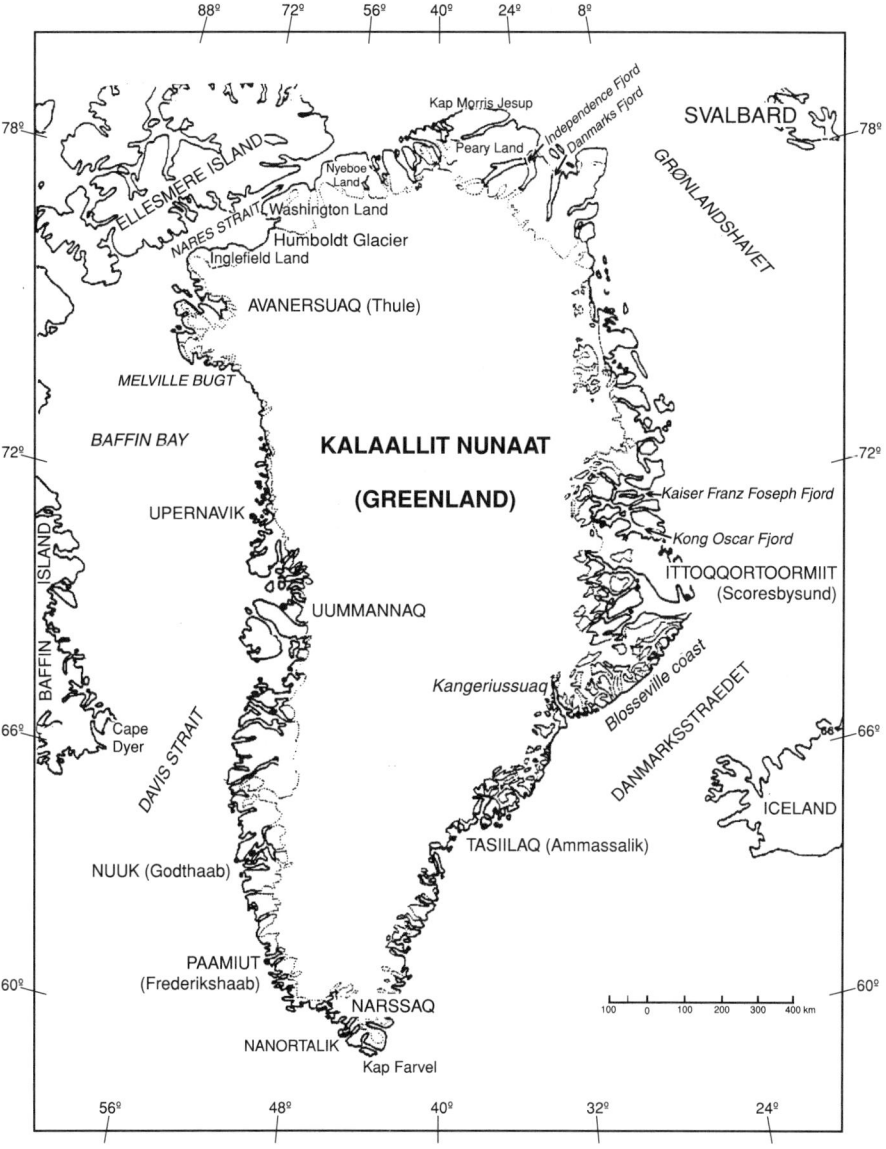

Figure 1. General Map of Greenland.

Introduction

Marc G. Stevenson
Canadian Circumpolar Institute
University of Alberta, Edmonton

Greenland's Inuit have for generations depended heavily upon the hunting and sharing of whales to fulfill their social, cultural, nutritional, economic, and spiritual needs. Yet, increasingly, the ability of Greenlanders to use whales in an ecologically responsible and sustainable manner is being threatened by individuals, groups, and nations who support animal rights over human rights and/or oppose the killing of whales for any reason. Whales have become a super symbol and 'cash cow' for many animal rights organizations as they maneuver to attack the heart and purse strings of the ill-informed. Regrettably, the 'anti-whaling industry' not only shapes public opinion, but influences the decisions of politicians and policy makers. Thus, Greenlanders and other community-based whaling societies live in a world of international whaling quotas and import restrictions on whale and other marine mammal products. Yet, at stake is not the fate of the whales — most whale populations are either stable or increasing annually in size — but the health, social vitality, cultural identity, and the community viability, of those small-scale societies worldwide that continue to depend on, and who care the most about, whales.

As a member of the International Whaling Commission (IWC), Greenland (Denmark) is subject to whaling quotas and other regulations set by the IWC. The IWC does not regulate the catch of small cetaceans, including beluga and narwhal, as the smaller toothed whales are not threatened from commercial hunting at present. Nor has the IWC placed any restrictions on the technology used by aboriginal whalers, although the Greenland Home Rule government continues to encourage more efficient and 'humane' killing methods. However, IWC regulations directly affect Greenlandic whaling as several species of baleen whales

Marc G. Stevenson

Table 1.
Documentation to the IWC on Greenland Whaling, 1979-95
(Source: Greenland Home Rule Government)*

1979

(Abor./Subs. Panel of Experts, Seattle)	*Subsistence Hunting: the Greenland Case.* (F.O. Kapel and R. Petersen).

1981

TC/33/WG/S3	*Subsistence Whaling in Greenland.* (R. Petersen, E. Lemche, and F.O. Kapel).

1983

TC/AB1	*Subsistence and Cultural Needs Relating to Aboriginal Subsistence Whaling Among the Inuit in Greenland.*
TC/AB2	*Nutritional Needs Relating to Aboriginal Subsistence Whaling Among the Inuit in Greenland.*

1984

TC/36/AS2	*The Greenland Aboriginal Whale Hunt: Report to the standing sub-committee on Aboriginal/ Subsistence Whaling of the International Whaling Commission, June 1984.*

1986

TC/38/AS3	*Documentation on the Catch Taken by Aboriginal People from the Central Stock of Minke Whales.*
TC/38/HK2B	*The Greenland Aboriginal Whale Hunt.* (P. Helms, O. Hertz, and F.O. Kapel).

1987

TC/39/AS1	*The Legal and Administrative Aspects of Whaling Operations in Greenland.*
TC/39/AS2	*Hunting Methods including the 'Cold/Warm Harpoon Question.'*
TC/39/AS3	*Scoresbysund: A Hunting Community in East Greenland.* (F.B. Larsen).
TC/39/AS 4	*Communal Aspects of Preparation for Whaling, the Hunt Itself and of the Ensuing Products.* (R. Petersen).

1988

IWC/TC/40/AS doc. 1	*Submission of the Delegation of Denmark.*
TC/40/AS 3	*Danish Statement.*
TC/40/HK 3	*Denmark's Answers to the Remaining Questions stated in Document IWC/39/19 'Report of the Humane Killing Working Group,' Annex 4.*
TC/40/HK 4	*Implementation of the Detonating Grenade Harpoon in Greenland's Whaling on an Experimental Basis.*

Table 1. (continued)
Documentation to the IWC on Greenland Whaling, 1979-95
(Source: Greenland Home Rule Government)*

1989
IWC/41/22	*Greenland Subsistence Hunting.* (J. Jervin, editor).
TC/41/HK 2	*Introduction of the Detonating Grenade Harpoon in Greenland Whaling on an Experimental Basis.*
TC/41/Inf 4	*National Inspection in Greenland.*

1990
TC/42/SEST 4	*Inuit and Whales at Sarfaq (Greenland): Case Study.* (S.E. Larsen and K.G. Hansen).
TC/42/SEST 5	*Cutter Hunting of Minke Whale in Qaqortaq (Greenland): Case Study.* (E. Josefsen).
TC/42/HK 1	*Introduction of the Detonating Grenade Harpoon in Greenland on an Experimental Basis.*
TC/42/HK 2	*Greenland Licences for Hunting Minke Whales with Rifles.*
TC/42/Inf 1	*Quota Monitoring in Greenland.*

1991
TC/43/AS 1	*Designation of Types of Rifles in Greenland.*
TC/43/AS 3 Add	*Conversion Factors for Minke Whale Meat* (Denmark).
TC/43/AS 4	*Qeqertarsuarmi arfanniarneq: Greenlandic Inuit Whaling in Qeqertarsuaq Kommune, West Greenland.* (R.A. Caulfield).
TC/43/HK 2	*Introduction of the Detonating Grenade Harpoon in Greenland, 1991.*
TC/43/Inf 1	*Quota Monitoring in Greenland, 1990.*

1992
IWC/44/HK 1	*Introduction of the Detonating Grenade Harpoon in Greenland, 1992.*
IWC/44/Inf. 1	*Quota Monitoring in Greenland, 1991.*
IWC/44/12	*International Register of Whaling Vessels, June 1992 (contribution concerning Greenlandic vessels).*

1993
IWC/45/HK 3	*Greenland Action Plan on Whale Hunting Methods, 1992.*
IWC/45/Inf. 1	*Quota Monitoring in Greenland, 1992.*

1994
IWC/46/AS1	*Whaling and Sustainability in Greenland.* (R.A. Caulfield).
IWC/46/AS2	*Quota Monitoring in Greenland, 1993.*
IWC/46/AS3	*Greenland Action Plan on Whale Hunting Methods, 1993.*

1995
IWC/47/Inf.2	*Quota Monitoring in Greenland, 1994.*

* Table does not include papers of the Scientific Committee.

occurring in Greenlandic waters (e.g., minke, fin, and humpback) fall under IWC jurisdiction. Although Greenlanders value the scientific advice of the IWC, they remain concerned that too much regulation from the outside, and too little local control over whaling, threaten the very foundation, fabric, and future of Greenlandic Inuit society. If there is a common theme that runs throughout the following papers, it is the articulate expression and validation of this concern.

Since 1979, Greenland has tabled numerous papers at IWC annual general meetings. Table 1 provides a partial list of papers presented to the IWC on Greenlandic whaling from 1979 to 1995. Many of these reports deal with aspects of quota monitoring and technological aspects of the whale hunt. However, many also illuminate and describe the economic, social, cultural, historical, nutritional, and spiritual importance and significance of whales and whaling to Greenlandic Inuit. It is these papers that constitute the contents of this book. The primary purpose of this collection is to make more widely available the valuable anthropological contributions made over the years by the Greenland Home Rule authorities, but not generally known outside meetings of the IWC. The papers have been heavily edited and, in some case, re-organized for clarity and consistency of presentation. Several photographs have also been added, while others have been dropped; suitable originals were not available at the time of printing. With the exception of Olsen's paper, contributions appear in chronological order so as to provide a historical view on the presentation of Greenlandic whaling issues before the IWC. A map of Greenland, locating the various whaling communities and locations referenced in the text, is presented in Figure 1.

Moses Olsen's paper, *Aboriginal Subsistence Whaling: A Contemporary Outlook* (1988), serves as worthy introduction to the volume. He makes the point, oft missed by industrialized nations in the south, that the use of whales in Greenlandic society is a form of respect and reverence, not of disrespect. Olsen sees the need for greater cooperation between local management systems and international authorities if this relationship is to be nurtured and sustained. Local management systems, and the ecological knowledge that informs them, are simply not sufficient to cope with the pressures (industrial contamination, global warming, etc.) experienced by Arctic fauna at the present time. While Greenlanders are committed to considering scientific advice and turning it into action, management decisions cannot be unilateral: Inuit must play a role in decision-making. Inuit must also begin to critically evaluate this scientific advice and data, and not accept them at face value, since authoritarianism and politicization often influence scientific decisions. Olsen subsequently exposes some of the questionable scientific assumptions that have lead to IWC regulations affecting Greenlandic whaling, and closes his paper with some intriguing suggestions about how Inuit and the IWC could work more closely together for the benefit of Greenlanders and the whales upon which they depend.

Subsistence Whaling in Greenland (1981) by R. Petersen, E. Lemche and F.O. Kapel describes the complex, multi-faceted nature of aboriginal subsistence whaling in Greenland. Aspects of Inuit involvement in the decision-making process and how IWC regulations are transformed into national policy are discussed. Whaling is viewed as an integral and necessary part of a broader complex of hunting and fishing activities by Greenlanders. While information about the occurrence and distribution of subsistence whaling and the technology used in the hunt is provided, the focus is clearly on the social, cultural, and nutritional benefits of subsistence whaling. The various uses of whale products are described, as is the important role they play in the contemporary Greenlandic barter and trade system. Also discussed is the vitally important role money plays in maintaining exchange parity, and keeping distribution channels open among hunters, fishers, wage-earners, and others with unequal access to whales. Additional sociocultural benefits of subsistence whaling as well as its nutritional contributions are enumerated, the latter most succinctly in an appendix by Dr. P. Helm, which clearly suggests that few other types of food available to Greenlanders are as nutritious as whale meat.

The development of a series of recommendations on how to improve management procedures to meet the needs of Greenlanders and IWC regulations is another major focus of this paper. Petersen *et al.* (1981) recommend a number of basic management principles for whales subject to IWC regulation and hunted by indigenous peoples. These include basing management decisions equally on the advice of the IWC Scientific Committee and on the documented social, cultural, and nutritional needs of indigenous peoples. Further, management measures should be restricted to quotas only, while Aboriginal whalers should be intimately involved in the decision-making process, allocating national quotas and applying their ecological knowledge to the estimation of catch levels. Petersen *et al.* (1981) consider the establishment of a number of processes and mechanisms that should allow Greenlanders to assume a larger active role in the management and monitoring of their whaling activities.

The Greenland Aboriginal Whale Hunt by P. Helms, O. Hertz, and F.O. Kapel (1984) evaluates the importance and roles of present day whaling in contemporary Greenlandic society. Many factors are discussed including:

- the current status of living resources,
- the occupational and demographic patterns of Greenland's population,
- the social and economic needs and conditions of Greenlanders, and,
- Greenlanders nutritional requirements.

Whaling in Greenland is seen to be a vitally important component of a broader system of subsistence and exchange. Particular emphasis is given to the socially integrative functions of whaling, to the cultural needs that whaling and the sharing of whales fulfil, and to the risks that imposed restrictions on Greenlandic whaling pose to cultural survival. Finally, the role of whale products as food in the community and their importance in the Greenlandic diet are discussed.

R. Petersen (1987) in *Communal Aspects of Preparing for Whaling, the Hunt Itself, and the Ensuing Products* describes the socially integrative functions of both whaling and the distribution of whale products. Whale meat and *muktuk* in Greenland are consumed domestically, shared locally, given away as gifts to more socially and geographically distant relations, and sold in local markets. While the first three uses tend to cement social relationships, the sale of whale products on the local market serves two diverse, but complementary, functions: 1) it allows wage earners and other people unable to partake in the hunt, to participate in the communal use of hunting products, and 2) provides a source of income to lower-income hunters that enables them to continue providing the community with fresh, healthy food. The fact that money has come to assume a significant role in the exchange and distribution of whale products cannot be denied. However, this clearly differs from large-scale commercial whaling in significant ways. In small economically-diverse Greenlandic communities, money allows families who would otherwise be unable to participate in the general distribution of whale meat to reciprocate within the framework of a mixed subsistence-wage labour economy. This contributes to a shared sense of cultural identity and community, as well as social equality, which prevents whaling from developing into anything different than what it has been for generations — a supplementary, yet significant, form of Greenlandic natural resource use motivated and sustained by socioeconomic, cultural, and dietary needs and values.

Scoresbysund: A Hunting Community in East Greenland (1987) by F. B. Larsen (with I. Egede and C.J. Jenkins) is a detailed examination of the roles of hunting and subsistence whaling in a remote Greenlandic settlement. Although modernization is a fact of life in Scoresbysund, it has supplemented rather than replaced the traditional hunting culture. With the collapse of the seal skin market, the economic situation of the hunter is such that wage labour is now necessary to maintain a lifestyle that no longer can maintain itself by hunting alone. Thus, the hunting economy is situated within a mixed (traditional and modern) economy. The various animals hunted and their uses provide a context for understanding the importance of minke whales in the local economy. Although minke whaling is a comparatively recent addition to the annual regime, it serves to stabilize the hunting-based economy. The contribution of minke whales to the general vitality of hunting as an occupation, and thus the health of the community, is especially important today when the hunting culture is threatened by external forces and a decline in subsistence production.

Greenland Subsistence Hunting (1989), assembled by J. Jervin (study coordinator) was produced for the IWC to comply with requests for updated information on Greenland society and hunting practices, and to illustrate the commitment of the Greenland Home Rule government to research and manage its living resources. However, only those sections of this document that relate to the economic, social, cultural, historical, nutritional, and associated values of whaling to Greenlanders are included in this volume. Definitions of traditional whaling are offered and the problems associated with the adoption of new technologies, such as the penthrite grenade, are described.

A factual description of the history and development of contemporary Greenlandic society is provided. In order to fully appreciate and understand the importance of subsistence hunting and whaling in Greenland today, background knowledge on various prey species and their history of use is presented. Numerous aspects of large-whale hunting in the past are also discussed, especially from the mid-1920s to 1959.

During this period, whale hunting was carried out by the Danish authorities to prevent hunger and malnutrition in Greenland's settlements, with the famous and cherished Greenlandic whale catcher, the *Sonja*, assuming a crucial role in saving Greenland's people from serious food shortages and malnutrition.

The valuable contribution that marine mammals make to the diet, health, and nutrition of Greenlanders is discussed from a historical as well as contemporary perspective, with P. Helms providing information on the modern situation. Although a much smaller percentage of the population today engage in hunting and whaling on a full-time basis, in the hunting districts these activities constitute a crucial and viable way of life that often escapes appropriate recognition in a statistical treatment of the national accounts.

The sociocultural role of subsistence hunting in Greenland, as well as key problems associated with this definition, are described by J. Dahl. He notes that marine-mammal hunting is neither a strategy of commercial enterprise nor a self-sufficient economy, but both — the distinction is arbitrary and meaningless in the Greenland context. He highlights the complex, integrative sociocultural functions of hunting, fishing, sealing, and whaling in the traditional mode of production, and the structural importance of subsistence activities in maintaining and reproducing means and relations of production from one generation to the next.

R. Petersen describes the traditional and present distribution channels governing subsistence hunting in Greenland. He illustrates why, in the absence of exchange parity following on the rapid modernization of the Greenlandic economy, the introduction of cash has functioned to keep game-sharing channels of distribution open, and thus the traditional bartering and sharing network alive. Sharing rules and customs are discussed, as are the roles that they, cash, and the commercialization of hunting play in maintaining social solidarity and a functioning economic system of balanced exchange.

Inuit and Whales at Sarfaq (Greenland) by S.E. Larsen and K.G. Hansen (1990) is an account of the Greenlandic municipality of Kangaatsiaq, with an emphasis on the settlement of Niaqornaarsuk. Today, the traditional subsistence economy of Niaqornaarsuk has given way to a mixed economy based upon fishing, wage labour, and hunting. The community is made up of numerous extended families whose members support and provide resources for each other — some by contributing products of the hunt, others by bringing money from wage labour positions into the household. Considerable discussion is devoted to: the social and symbolic importance of consuming and sharing whale meat and *muktuk*, and the personal development of the individual's independence *vis-à-vis* obligations to other extended family members and one's role in the larger community. The problems associated with the hierarchically-structured methods of wage labour employment are outlined together with the mitigating collective subsistence activities and social events which serve to bind the community together.

Against this backdrop, a collective whale hunt using small boats is described, with special emphasis on how it develops, who becomes involved in the hunt, and how the hunt is conducted. Also presented is a description of a cutter whale hunt, which, in contrast to the opportunistic nature of collective (small-boat) whaling, is usually planned in advance to take a 'quota' whale. While the flensing process is described for both types of whaling, discussion is centered around how the catch is distributed to various participants in the hunting and flensing operations. Associated with collective whaling are three levels of distribution, each with its own social and symbolic significance. In contrast, there are only two levels of distribution in cutter whaling, one of which involves the sale of whale products in open air markets — an important means by which most Greenlanders, who have no opportunity to participate in whaling, can obtain whale products. An analysis of households involved in whaling activities as well as the utilization, distribution, and storage of whale products illustrates how whaling is both an integrated and an integrating feature of the economic, social, and cultural life of the settlement. While termination of collective and cutter whaling will not devastate the local economy to the point of collapse, eliminating whales from an already limited and balanced set of options threatens the very foundations of local existence.

E. Josefsen's (1990) *Cutter Hunting of Minke Whale in Qaqortoq (Greenland)* is another detailed case study outlining the economic, social, cultural, and nutritional importance of contemporary cutter whaling to a small Greenlandic community. The crew selection, composition, and division of labour on two cutter whalers in Qaqortoq are described. The cutter whaling season is discussed in the context of fishing and other seasonal subsistence activities undertaken by fishers and hunters in Qaqortoq. Cutter whaling involves several integrated stages, including preparation for the hunt, the hunt itself, flensing, distribution of whale products (including sale in local markets), and the use and consumption of whale products. These, as they relate to the actual hunting and distribution of a whale, are described at length. Included as an appendix is a history of cutter whaling in Greenland, with an emphasis on the *Sonja*, a boat of considerable historic and symbolic significance to Greenlanders.

Richard Caulfield's *Greenland Inuit Whaling in the Qeqertarsuaq Kommune* has been extracted from the author's *Qeqertarsuarmi arfanniarneq: Greenlandic Inuit Whaling in Qeqertarsuaq Kommune, West Greenland* (1991). An extensively revised version of this work is in press (Caulfield 1997). Although *Qeqertarsuarmi arfanniarneq* is a comprehensive account of the meaning and significance of whales and whaling to the Inuit of the Qeqertarsuaq Kommune, only two chapters of that report — one describing the nutritional and sociocultural significance of minke and fin whaling, the other analyzing contemporary whaling issues in Qeqertarsuaq Kommune — have been included in this volume. While whale meat, *mattak*, blubber, and the highly-prized belly flesh are used for household consumption, in sharing networks between households, and for sale to others, cash is now integrated in the network of whale meat exchange and distribution. Two types of distribution are noted: one associated with collective whaling, whereby whale products are distributed equally to all those present during the activities; the other with whaling from fishing vessels, whereby the whale is divided between the vessel itself (for expenses incurred), the vessel's owner and operator, and the crew members. Typically, whale products from such hunts are sold on the local market to people who have no other way to obtain whale products. Although avenues exist for selling whale products, current whaling quotas are not sufficient to meet the local demand. Thus, less than 2% of all households surveyed in 1989 sold such products.

The consumption and sharing of Greenlandic wild foods, or *kalaalimerngit*, form an integral part of local diets while strengthening cultural identity and social solidarity. Caulfield's household food consumption survey data reveal that marine mammals, especially seal and whale, constitute a major portion of the diet. These foods, in comparison with *qallunaamerngit* (white man's food), provide a low-cost and readily available safety-net during periods of unemployment, and are integral to sharing networks which reinforce Greenlandic Inuit identity and community solidarity. Sharing networks based upon gifts of meat, *mattak*, and fish, bind extended families and strengthen the sense of community. At least 50% of all households surveyed regularly share whale meat and other wild foods with immediate and extended family members, relatives in other communities, and friends and acquaintances. While sharing between households has decreased in recent decades, because increasing amounts of meat and fish are sold for much needed cash, the procurement, preparation, sharing, and consumption of whale products and other wild foods provide a sense of continuity with Inuit cultural traditions, while integrating families and communities. Furthermore, whaling has begun to serve as a rallying point for the expression of Greenlandic national identity as hunters increasingly find their use of marine mammals being challenged by non-Inuit interests at the international level.

In recent years, critics have asserted that Greenlandic whaling is increasingly being driven by profit maximization, capital intensification, and commoditization. Caulfield directly challenges these assumptions by demonstrating that neither fishing-vessel whaling nor collective whaling in the Qeqertarsuaq Kommune are carried out to maximize profit. Whaling, though important for many other reasons, is an economically marginal activity; gross income from fishing-vessel whaling does not typically exceed 10% of the total annual income for these boats, and cash plays an even smaller role in collective whaling where culturally-based egalitarian practices normally leave individual hunters with only a small share of the catch. Moreover, cultural factors continue to negate efforts to maximize profits from whaling. Although capital intensification is occurring in Qeqertarsuaq's fishing fleet, it is principally associated with the primary income-generating activities (i.e., commercial shrimping and fishing), rather than whaling. As Caulfield points out, new investments in whaling technology come not from economic motivation at the local level, but from demands at the international level for more 'humane' killing of whales. External (IWC) regulation of whaling has significantly reduced the number of whales caught, placing a strain on culturally-based hunting practices and increasing the alienation of hunters from each other as well as from whaling management regimes. In order to satisfy these international demands, the Home Rule government limits whaling to only those who are registered as full-time hunters or fishers. This regulation, while designed to protect traditional livelihoods, excludes many part-time hunters who previously had a long history of whaling, thus expanding social differentiation and the often negative consequences identified above.

Caulfield's (1994) *Whaling and Sustainability in Greenland* evaluates the 'sustainability' of Greenlandic whaling using a method developed by social scientists (Young *et al.* 1994) for identifying whether or not societies practise sustainable and equitable resource use. Caulfield demonstrates that Greenlandic whaling is carried out by social groups sharing a common culture whose maintenance depends upon whaling and/or the consumption of whale products. Thus whaling, while playing a central role in maintaining social relations within the group, serves to reinforce social and cultural norms. As Greenlandic whaling is confined to operations associated with specific shore-based whaling communities, open access to marine resources is restricted and stewardship and effective management practices are promoted. He shows that the process of handing down knowledge from generation to generation serves to maintain appropriate Inuit-whale relationships, while reinforcing social relationships based on kinship or other alliance-creating traditional institutions. A key difference between sustainable and unsustainable whaling is whether or not whaling has values beyond the purely economic; whaling for Greenlanders clearly has social, cultural, nutritional, spiritual, and other values not directly related to economics. Although these criteria may be necessary for sustainable and equitable whaling, there must be some mechanisms to promote the biological sustainability of the resource. In this regard, Caulfield demonstrates that the Home Rule government's regime for managing whaling is increasingly effective and responsive to ensure the biological sustainability of Greenlandic whaling. Caulfield concludes that in order for whaling to be a major contributor to sustainable development in Greenland, there must be greater understanding of mixed subsistence-cash economies, more effective management and monitoring regimes, and greater recognition and understanding of the diversity of human adaptations.

Collectively, the papers in this volume demonstrate the very high level of commitment that Greenlandic authorities have shown towards conserving whales and sustaining the Inuit communities that depend on them. There is no contradiction between use and conservation. This is a false dichotomy set up by the industrialized south which, up to this point, has only exhibited an exploitive, unsustainable relationship with nature. Where human beings see themselves as an integral part of natural ecosystems, use is and can be an excellent conservation strategy. For example, both Inuit and scientists have recognized that whale populations that are hunted on a sustainable basis have less disease, more food, and reproduce faster than whale populations that are not hunted sustainably (Department of Fisheries and Oceans, Canada, 1994). Also, Inuit use of whales entails a level of respect and caring, which includes the continual cost-effective acquisition and refinement of ecological knowledge about whales that serves the best interests of whales. Who will be in an effective position to conserve whales, other living Arctic resources, and the Arctic environment if the use of these resources is no longer permitted? Arguably, the imposition of unnecessary restrictive regulations that significantly impede the use of whales by Greenlanders and other small community-based whaling peoples, threatens not only these communities' economic and cultural survival, but also threatens the long-term survival of the whales themselves.

As a group, the papers in this volume illustrate the level of concern and responsible reaction by Greenlanders toward increasing efforts by others to exert external control over their local whaling activities. Perhaps more than anything else, these papers argue for greater and more meaningful Inuit participation in decision-making, e.g., by contributing knowledge and understanding and developing regulations that accord with local realities. Only through increased local control over managing and monitoring their whaling activities will Greenlanders affect their own well-being and that of the whales and marine ecosystems of which they are an integral part.

Bibliography

Caulfield, R.A. 1997. *Greenlanders, Whales, and Whaling: Sustainability and Self-Determination.* University Press of New England, Hanover, NH.

Department of Fisheries and Oceans. 1994. *Southeast Baffin Beluga Co-Management Plan.* Prepared for the Minister, Ottawa.

Young, O.R., M.M.R. Freeman, G. Osherenko, R.R. Anderson, R.A. Caulfield, R.L. Friedheim, S.J. Langdon, M. Ris, and P.J. Usher. 1994. Subsistence, Sustainability, and Sea Mammals: Reconstructing the International Whaling Regime. *Ocean & Coastal Management* 23:117-127.

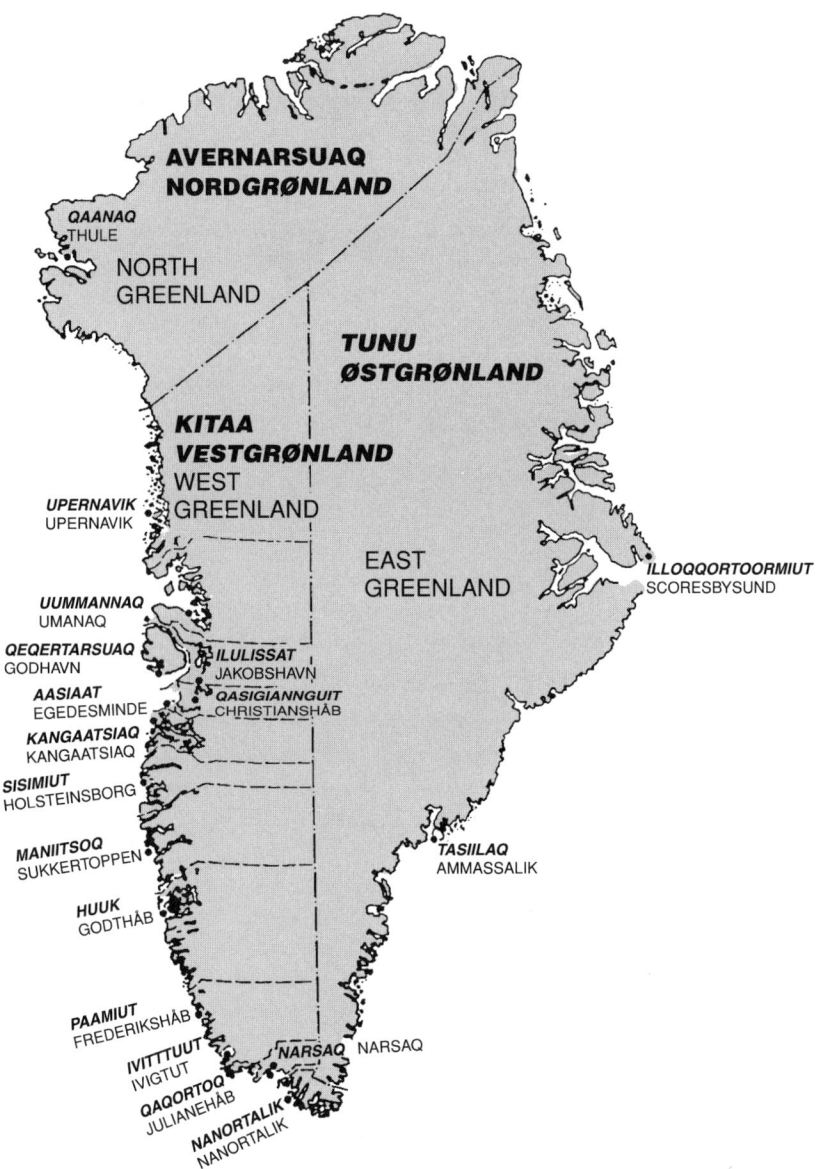

Figure 1. Community-based Whaling in Greenland.

Aboriginal Subsistence Whaling

A Contemporary Outlook

Moses Olsen
Minister for Fisheries and Hunting,
Greenland Home Rule Government
1988

> The whale is a split in our consciousness: on the one hand viewed as a product, as a resource, as an article, an object to be carved up to satisfy the economic imperatives; on the other, a view almost lost now, as the great leviathan, the guardian of the sea's unutterable mysteries.
> (John McIntyre, *Mind in the Waters,* 1974:8)

Aboriginal whalers do not suffer from this split, so characteristic of modern civilization. To the Inuit hunter, the whale is a mysterious and wonderful creature. Therefore, from icy waters, seemingly so sterile, the whale emerges with his unbelievable bulk of flesh and fat. Placing itself, the whale saves the hunter and his family, indeed the entire village, from malnutrition and misery.

The whale, along with the polar bear, is the most respected of all prey. Our forefathers used to dress up, as for a wedding, when they went out for the nerve-racking adventure of small boat whaling. To this day, in Alaska, the bowhead whale hunt takes on religious significance. The whale as food, as a 'resource,' is to us in no way opposed to the reverence we feel for this stupendous creature. On the contrary!

In modern jargon one could say that our perception of the whale is 'holistic.' All aspects of the whale fit together in a wonderful whole. The whale is beautiful, tremendous, awe-inspiring and tastes good. He simply represents one of the Creator's most amazing creatures.

We wouldn't want to be without him.

Nutrition

The starting point, the basic issue, the substance of the master and bottom line, all come down to one and the same thing: food.

For millennia whale meat has ranked among the top species in the Inuit diet, from the Bering Strait across the North American Arctic to the east coast of Greenland. It is not difficult to see why. Whales, as prey, range from large to immense. One single animal will fill many empty stomachs. Whale meat and whale blubber is healthy. The animal fats they contain help lower the cholesterol level in the human blood stream, they prevent coronary diseases and they favor longevity. Whale skin, *muktuk* in the Inuit language, is a precious delicacy with its distinctive nutty taste. It is digested slowly and contributes vitamin C as well as a lot of calories to the hunter on his endless trek through the icy landscape. The baleen, or the teeth, according to what kind of whale has been taken, have immediate or exchange value. Even the bones, from sizable to huge, have found and continue to find uses. Nothing is wasted, and what is not edible, such as bone and baleen, is turned into handicrafts.

Inuit have never hunted any whale species to extinction, not even close. Several factors account for this:

1) Inuit are not a numerous people (only approx. 100,000), but their hunting grounds, which spread out contiguously from Siberia to East Greenland, cover an immense area – about ten times the size of France. Hunting pressure on the various types of wildlife is laid out thin, so to speak.
2) Traditionally, hunting pressure Is spread over many different species, according to seasonal changes, and the geographical proximity of various migration routes.
3) The Inuit homeland is 'whale country' par excellence. Whales abound, less now than in the old times, before commercial whaling of the whiteman, but there are still very many.

Aboriginal Subsistence Whaling

The Need for Management of Whale 'Stocks'

All in all, one may say that aboriginal subsistence whaling across the Arctic is alive and well, and so are the whales, according to the judgment of the hunters themselves.

All the same, several factors have clearly necessitated a new kind of whale management on the local level. With the introduction, in the 1950's, of small motorized fishing vessels with harpoon canons, hunting techniques became more efficient. Nowadays, no utilization of whales can take place without some kind of supervision and management.

In the old days, rational wildlife management by state authorities was by and large unnecessary, if for no other reason than most of the Inuit were migratory for the better part of the year. 'Wildlife management' in those days consisted, in part, simply of letting vast hunting grounds alone for long periods of time. Of course, this, the best of all types of management, has become impossible in modern times. All kinds of new pressures and requirements have forced people to abandon nomadic lifestyles, and modern vehicles and vessels move around at speeds and distances hitherto unknown.

Formal wildlife management of some kind is a recognized requirement nowadays. This is clearly acknowledged by contemporary Inuit. But the Inuit hunters expect, just as clearly, to influence management measures as well as enforcement rules. After all, the measures considered take place in their country, and it is their livelihood which is at stake.

The Need for International Advice on Management

Obviously, traditional hunting communities possess a wealth of detailed knowledge of wildlife in the midst of which they live, and which constitutes their primary source of food. Just as obviously, these insights are the backbone of present Inuit wildlife management at the local level. Local management sets rules of conduct and conservation measures, which vary greatly from one end of the Arctic to the other, but which flow from the age-old symbiosis between hunter and prey. No hunter in this historic tradition has ever been motivated by personal gain to the detriment of any species.

However, with the unprecedented pressure brought upon contemporary wildlife by modern man, one may reasonably ask whether traditional insight and local measures are alone sufficient. Can local management systems prevent the spread into the Arctic of the global degradation of the natural environment originating in countries having adopted the industrial way of life? Many, including those in the hunting communities of the North, would agree that they can't and that we have reached the point where international cooperation has become necessary.

For centuries, Inuit communities have witnessed the over exploitation of the great whales. The bowhead, the blue whale, the sperm whale, the humpback, once plentiful in the vast straits and inlets of the Inuit homeland, were all at one point reduced to near extinction by whalers of Dutch, British, and American descent, an enterprise motivated by greed and profit, inexplicable and distasteful to the Inuit. The creation of an international regulatory body like the International Whaling Commission (IWC) is perfectly legitimate and deserves Inuit support and cooperation. Inuit, more than any others, are glad to see Arctic whales recovering the way they are. Inuit support the moratorium in commercial whaling. Let there be no doubt about that.

This doesn't mean that Inuit agree to every stock assessment, scientific procedure or quota recommendation that proceeds from the IWC. Far from it. The application of exceedingly rigorous and complicated arithmetical calculation models to the minke whale population of the Davis Strait, for example, is built on very dubious assumptions rather than on facts. This has resulted in a more than 50% reduction of the allowable catch, and has stunned and angered the few thousand people who depend on whales for their subsistence. In Alaska, the Alaskan Eskimo Whaling Commission (AEWC) has had a long and dramatic history of disagreeing with official stock assessments of the Bering Strait/Beaufort Sea bowhead (previous official stock assessments have now turned out to be wrong).

Inuit far from agree on everything coming out of the IWC. But then Inuit, be they the Inupiat of Alaska, the Greenland Home Rule authorities, the Fishers' and Hunters' Association of Greenland, or the Inuit Circumpolar Conference, have clearly accepted the existence of the IWC and want to cooperate. Inuit know that there exists a need for international scientific advice on the sustainable use and conservation of whales. Inuit are not averse to unbiased advice.

But, and this is the crux of the matter, Inuit are averse to having rules and regulations imposed on them. Colonial times are over. No distant government, let along any lofty international body, is in a position to tell Inuit what to do. This day and age calls for consultations, dialogue and cooperation in management decisions. It is in the interest of suitable whale stock management that aboriginal subsistence whalers be part of the decision making process, that their voice be heard, and that their needs and reasonable concerns be addressed.

Using Advice in the Formation of Management Decisions

Using scientific advice in the formation of practical management decisions is, in the international community, accepted as ordinary civilized behaviour. This kind of behaviour is expected also of aboriginal subsistence hunters. The organizations of aboriginal subsistence hunters, non-government organizations, and, in the case of Greenland, the autonomous government, have clearly signaled their intent to become involved in the making of wildlife management decisions. They too are committed to considering scientific advice and turning it into action, when needed and necessary. However, for this to occur two conditions will have to be met:

1) the scientific advice has to be solid, and;
2) aboriginal subsistence whalers have to be involved in the decision-making process.

> *'The scientific advice in question has, in actual fact, to be solid'*

Is all of the scientific advice coming from the IWC really solid? As a non-scientist myself, I have my doubts, and I am not alone in my skepticism. Laymen often find themselves in the situation where they have to take a scientist's word for the authenticity and validity of a certain 'fact.' To many people in this day and age, the natural sciences carry the exalted aura of a religion about them. Great scientists are sometimes treated as high priests. They command respect and receive it. This, of course, risks fostering an unsound authoritarianism.

Politicization is one danger threatening the scientific work being done today and so is the unhappy tendency to operate with uncertainties as if they were facts. Lay people must have a way to test the authenticity of a given scientific statement. We too must be critical. And when it comes to corporate scientific statements issued by a multinational body of scientists, the test is easy: true scientists do not, as a rule, arrive at precisely the same conclusions as do political decision-makers of their

Moses Olsen

respective national governments. The political process is one thing. It is ordered by a will, proceeding from a government which has certain objectives and priorities, dictated by the necessity of its own system. The scientific process, on the other hand, is something completely different. It is cognitive and tries dispassionately and without any preconceived bias, to formulate the correct answer to a given question.

It may so happen that the answer arrived at by a scientist turns out to support the political position of her or his government. But it also may turn out that it doesn't. Nobody wants to support overtly the idea of the scientific conclusions being dictated beforehand by the national government of the scientist in question.

The application of a whale stock calculation formula, for the assessment of the Davis Strait minke population, is as follows:

$$fN_i = (fN_{i-1} - fC_{i-1})S + r_{i-t} \times fN_{i-t}$$
$$mN_i = (mN_{i-1} - mC_{i-1})S + r_{i-t} \times fN_{i-t}$$
$$r = M(1 + A(1 - (N_{1-t/No})^n))$$
$$N_{i-t} + fN_{i-t} + mN_{i-t}$$
$$fN_{o=0}$$

This is one feature of the work of the IWC, which makes Inuit suspicious, to say the least. Although this formula is fraught with uncertainties, it nevertheless is allowed to result in quota cut-backs of no uncertain character. This formula neatly skips over the facts that the assumption of the existence of two separate minke whale populations in Greenland waters, one along the west coast and one along the east coast, is completely arbitrary. Scientists have drawn a line from Cape Farewell and due south, making their calculations separately to the west and to the east of that line, respectively.

Also, in this stock assessment exercise, nobody has any idea of the natural mortality rate of the minke whale. Minke reproduce every year, but nobody knows: a) whether there is any difference in the spontaneous infant mortality rate of male and female offspring; b) about the effect of killer whale predation, or; c) whether there is any difference in the fertility of the (supposedly) different stocks of the North Atlantic.

Likewise, nobody has any exact idea of the recruitment rate of these (supposedly separate) stocks. Is it 2%, 4%, …? These figures make quite a difference when determining a sustainable 'harvest' level.

The most serious misgiving of all, when dealing with the scientific work of the IWC, has to do with the way the 'findings', oft times based as they are on a number of assumptions, are dished out to the public. They are presented as scientific 'facts' which patently they are not. Thus, an animal protection spokesman feels justified to tell the Danish public that 'it has been scientifically proven' that Greenlanders ought not take more than 50 minke whales per year. Pure nonsense, and certainly scientifically unproven. Why, then, is there no protest from any IWC scientist when their work is being thus misrepresented?

> *'The aboriginal subsistence whalers themselves have to be involved in the decision-making process'*

We are faced with a requirement which, till this day, may not be so well understood. It is nevertheless called for, for at least two reasons. Common decency requires local participation in the management debate and ensuing decision-making. People whose livelihood is at stake, and who are the exponents of age-old cultural traditions relevant to the issue in question must participate in decision-making. Moreover, without the participation of the aboriginal subsistence hunters themselves, the IWC risks adopting management policies that are impractical, shortsighted, or impossible to implement. Aboriginal whalers are conservationists themselves and do have contributions to make toward a sound set of management rules. There is no sense in wasting this potential.

Positive Thinking

Aboriginal subsistence whalers have a role to play in the IWC and international whalers have a role to play in the IWC. The international whale research community has no better situated allies. If the IWC could advise the Inuit on a whale research design that is practical and suited to the resources of the local communities, Inuit may gladly cooperate. There is no doubt that Inuit organizations would welcome IWC funds to carry out research that is beyond the capacity of local resources.

As to the question of how this research could be turned into management decisions, that, of course, is an extremely sensitive issue. But Inuit are not closed to dialogue. What is imperative is that their personal experience as whalers as well as the accumulated knowledge and insights of centuries of whaling be acknowledged for what they are, and that Inuit are respected as equal partners in the management decision process. The issue is to find a mechanism for aboriginal participation in this process. If Inuit are treated in an authoritarian or paternalistic manner they'll close off, and follow their own ways, in the immense and unenforceable hunting grounds of the Arctic. This is not in the interest of the IWC.

As the IWC is run today, it is hardly equipped — maybe not even minded — to deal with the aboriginal subsistence whaling issue in an equitable manner. That must be changed. The IWC could invite and financially support aboriginal subsistence whalers from Greenland, Alaska, St. Vincent, and the Grenadines to partake as equal partners in the work of the Aboriginal Subsistence Whaling Committee. Let deliberations aimed at rule-setting for the great whales of the Arctic take place in the Arctic, thereby showing a little bit of respect for the people and the livelihood of those the IWC is trying to infringe upon. Create an IWC foundation for 'The Ice-Floe People,' in order to institute an Arctic whale research project with the express purpose of capitalizing on ecological and traditional Inuit knowledge of whales, so that this treasure of insight is recorded and made available for the outside world.

In the interest of elementary democracy, and in the interest of developing comprehensive and workable management plans for subsistence use of wildlife, including whales, it is essential that the aboriginal people be fully involved at all steps of development and implementation of management programs.

Please do understand that when faced with aboriginal subsistence whalers, you are dealing with a conservationist culture. It cannot possibly be the goal of IWC to take away the Inuit traditional capacity to manage and conserve. Inuit do have something to offer. Inuit expect to be called upon to contribute their piece.

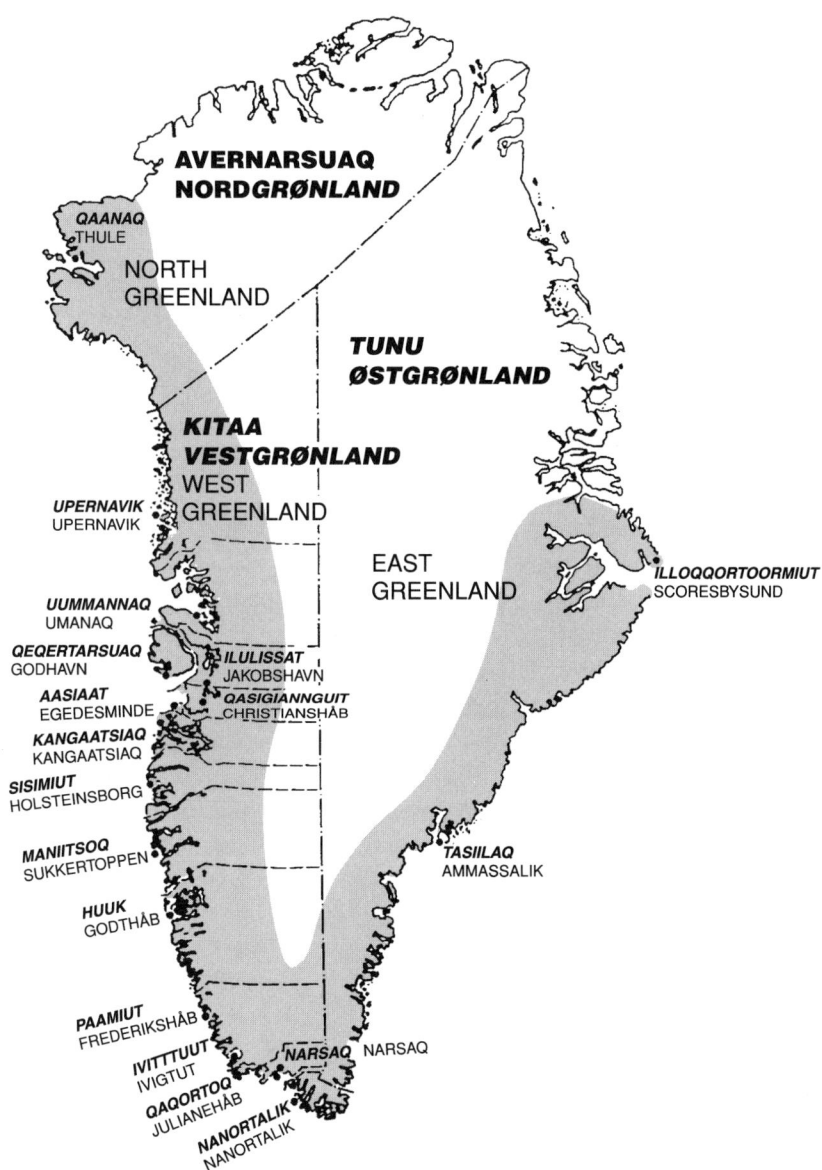

The Main Community-based Whaling Districts in Greenland.

Subsistence Whaling In Greenland

R. Petersen
Professor, Institute of Eskimology
University of Copenhagen

E. Lemke
Danish Commissioner to the IWC
Ministry of Greenland, Copenhagen

F.O. Kapel
Greenland Fisheries Investigations
Charlottenlund, Denmark
1981

Introduction

In Greenland, the concept of regulating hunting on a local basis has roots in traditions more than a century old. Formerly, local councils in Greenland had advisory authority, but in reality hunting regulations were established by these councils. The Home Rule System in Greenland is relatively new. However, more legal competence is gradually being assumed by the Home Rule Authorities. This is also true within the framework of international agreements to which Denmark is a party.

The indigenous peoples of Greenland are represented by the Greenland Home Rule Authorities or Grønlands Hjemmestyre. The Hjemmestyre participates in the Danish decision-making process before and during the meetings in IWC, through its representative in the Danish delegation. Following the IWC-meetings, the decisions of the Commission, which are relevant to Greenland, are submitted for consideration by the Hjemmestyre. The Hjemmestyre may eventually request that Denmark oppose decisions which are considered unacceptable for the Greenland community.

Before the Danish Government will issue regulations on whaling in Greenland, which normally is done only with the aim of transforming IWC-decisions into national legislation, the draft regulations are submitted for consideration by the Hjemmestyre following the normal hearing procedure between the Government and the Hjemmestyre. Aspects of involvement of indigenous peoples in decision-making processes are dealt with in greater detail below.

General Information on Greenlandic Whaling

Occurrence and Distribution of Subsistence Whaling by Indigenous Peoples

Subsistence whaling in Greenland is an integral part of a complex pattern of hunting and fishing activities by Greenlanders, a point of view documented in a paper presented at the Panel Meeting of Experts on Aboriginal/Subsistence Whaling held in Seattle in February of 1979 (Kapel and Petersen 1979). This paper contains much of the information pertinent to the occurrence and distribution of subsistence whaling in Greenland.

Areas in Which Subsistence Whaling Occurs

Hunting of whales occurs in all inhabited parts of Greenland. The relative importance in different regions is reviewed in Kapel and Petersen (1979:4-6). More detailed information on the regional and seasonal distribution of the hunting of the various species may be found in several papers previously presented to the IWC (Kapel 1977b, 1977c, 1978a, 1979a, 1980, and Danish Progress Reports for 1976-77 to 1979-80).

Species Taken and 'Stocks' from Which Takes Occur

A complete list of the cetacean species taken is found in Kapel and Petersen (1979: 9-12) where comments on 'stock' relationships are also given.

Methods and Equipment Used (Past and Current)

Detailed accounts of methods and equipment used for hunting in Greenland are available since the time when the Danish-Norwegian colonization started (e.g., Egede 1741 [1925]; Fabricius 1818a, 1818b; Rink 1852, 1857, 1877 [1974]; Amdrup *et al.* 1921; Hansen 1971 [1922]. Modifications and innovations have taken place all throughout this period. Some elements of present modifications and current hunting methods are referred to in Kapel and Petersen (1979: 11-12, 30-37) and in Kapel (1975c, 1977b, 1978a, and 1979a).

To summarize, small cetaceans are hunted from small powered boats — in some areas in combination with, or exclusively from kayaks — using hand harpoons and rifles. Drive hunting methods, where several boats cooperate, are sometimes used. Narwhal and beluga are also hunted at the floe edge or along leads in the spring. Minke whales are taken mostly by small fishing vessels equipped with harpoon cannons using non-explosive harpoons for fastening, and large calibre rifles for killing, if necessary. In some areas minke whales are hunted by several small boats in cooperation, using hand harpoons and high-power rifles (cf. Kapel 1978a). Fin and humpback whales are only taken by some of the larger harpoon-vessels, employing the same method used in hunting minke whales.

Catch: Historic and Recent

Information on past and present catch levels is summarized in Kapel and Petersen (1979: 9-12, Fig. 5-7), and presented in greater detail in Kapel (1975c, 1977b, 1977c, 1978a and 1979). Data on catch levels for the Greenlanders' whale hunting prior to 1900 are incomplete or non-existent, and indirect evidence is anecdotal, or have not been treated in detail in the above mentioned papers.

Commercial Catches from 'Stocks' where Subsistence Whaling Occurs

Whaling for bowhead whale in Davis Strait by various European countries in previous centuries is considered one of the main causes for the depletion of this population. This has led to a cessation of a subsistence take of this species in Greenland.

Commercial whaling for humpback whale in Davis Strait, and in other areas of the North Atlantic, in the early part of the 20th century probably contributed greatly to a reduction of this species (Kapel

1979:198, 213). A small number of humpback whales (avg. = 4, max. = 12 per year) was taken by a modern catcher boat operated by the Royal Greenland Trade Department between 1924 and 1953. In the same period, subsistence whaling of this species by Greenlanders was discontinued, and the meat of the humpbacks was used exclusively for local consumption.

Similarly, the meat of the fin whales (avg. = 22, max. = 53 per year) taken by the same boat in the periods 1924-39 and 1946-58 was used locally. The operation was semi-commercial as only the blubber was used for industrial purposes. In the periods 1919-25 and 1931-37, however, Norwegian pelagic fleets operated in Davis Strait, taking more than 500 fin whales. Combined with the Greenland catch, this take may have had some effect on the size of the population (Kapel 1979). Since 1958, commercial catches of fin whales have not been taken in Davis Strait.

In 1968 Norwegian vessels began catching minke whales off West Greenland, taking an average of 170 animals per year until 1976. Since 1977, Norwegian activity in that area has been limited to one vessel taking an annual quota of 75 minke whales per year.

Products and Processing of Subsistence Catches by Indigenous Peoples

Some of the issues relevant to an evaluation of present day use and importance of whale products in Greenland are dealt with in *Subsistence Hunting — The Greenland Case* (Kapel and Petersen 1979: 23-29, 37-40). Nonetheless, additional comments are given below.

Products: Processing and Use

Since ancient time the most important whaling products were meat, *muktuk* (whale skin), blubber, baleen, and teeth. As mentioned elsewhere, whaling was scarcely the only or most stable basis for subsistence in ancient Greenland. Nevertheless, an occasional and opportunistic catch of whales was of great importance.

In the Greenlandic communities of old, living resources were regarded as common heritage and collective property. Individual households could, however, take supplies for private use according to certain rules (e.g., when a household had reserved some parts for itself, other members of the community would loose their rights to that part of the common property). A captured whale belonged to the common resources of the community and everybody from the settlement and even neighbouring settlements could take meat, *muktuk*, blubber, baleen, or teeth, which then became property of the individual households. The meat was used for human consumption and in the northern parts of Greenland, where the dog sled was a vital part of the hunting culture, for animal fodder as well. The *muktuk* was highly prized and reserved for human consumption.

Blubber was consumed, but mainly it was utilized for lamp fuel in lamps for lighting, heating and cooking. In winter, the blubber could be of even greater importance than the meat as serious consequences would be experienced if the lamps had to be extinguished.

Baleen was used for various household tools, probably the most important of which was the manufacture of lines for fishing deep sea species such as redfish (Sebastes) and Greenland halibut (Reinhardtius) from the ice (Egede 1741:1770). This fishery was especially important for those who were unable to hunt seals. Teeth as well as parts of the skeleton were used for harpoon heads and for equipment and parts associated with the kayak and dog sled.

This utilization of whale products certainly meant that the catch of even a single whale contributed significantly to the economy of the community. What made whaling different from other more routine hunting activity was (and is), however, its profound socioeconomic importance. An abundance of whale products contributed towards equalization of social and economic differences between households. Those families without a provider/supporter were, like more fortunate families, able to obtain important shares of meat, blubber and baleen. When a whale was flensed far from the settlement, orphans were allowed to take a share from the products brought home (Glahn 1921:135).

After the colonization of Greenland in the early part of the 18th century most of the baleen and blubber was sold to the Trade Department or bartered for European goods. Whale and seal blubber remained an important Greenlandic export product well into the 20th century. After World War II, however, the market collapsed and the industry was closed down in the 1950s. In recent years, an increased interest in making refined products from whale and seal blubber has emerged in Greenland, and the possibilities are being investigated.

The meat and *muktuk*, however, remain important contributions to local subsistence in Greenland. During the semi-commercial interlude of modern whaling, operated by the Royal Greenland Trade Department in the early part of the 20th century, only the blubber was exported, whereas the meat and *muktuk* were disposed of locally, either freely or at very low payment (Kapel 1979:198). Owing to Greenland's rapidly increasing population, the present demand for the edible whale products is also growing and the local market is far from satisfied (Kapel and Petersen 1979:23-29).

Nature and Level of Distribution, Barter, and Trade

Since ancient times Greenlanders from different areas have met in specific places where possibilities for seasonal hunting were optimal. These places also functioned as markets where trading took place. Baleen and fishing lines made of baleen were important barter objects on those occasions, because the catch of great whales mainly took place in certain areas, e.g., Disko Bay and Sisimiut (Holsteinsborg), and for humpback whales, Paamiut (Frederikshab)(Rovsing 1976:175). Food products from whales were apparently not bartered in the same manner, but dried meat, *muktuk,* and flippers were often consumed when entertaining visitors from other settlements.

Thus, barter was necessary, even in an egalitarian society, when the geographical distribution of resources was uneven. Sharing of food and entertaining visitors was, of course, part of everyday hospitality. In the long run exchange through barter and sharing could perhaps be considered as a common form of mutual insurance against times of misfortune or need. This kind of mutual insurance could best take place in a homogeneous community where provisions could be bartered for other provisions.

When a community becomes less homogeneous, for whatever reasons, the conventional distribution channels may be disrupted because some members of the community may still be interested in receiving certain provisions, but are unable to pay in kind. Such a development has taken place in Greenland, and in certain local communities, attempts to limit the distribution channels to groups with equal possibilities have been tried, e.g., when the production of country food between fishermen and seal hunters was uneven (Kleivan 1964:65). In other communities, it was found that distribution channels could be retained if those who were not able to pay in kind could pay in money. Significantly, the introduction of money in this manner into the distribution system did not change the conventional patterns of utilization of the living resources. The distribution channels remained, on the whole unchanged, and the exploitation of the resources was held at the same level. Maintaining the distribution channels also meant that those fishermen wanting to obtain products of the hunt were not motivated to spend their time hunting.

In general, today there are three categories of hunting licenses for resident Greenlanders:

1. for persons whose main occupation is hunting or fishing;
2. for persons to whom hunting or fishing is a secondary occupation, but a necessary supplement to their lifestyle; and,
3. for persons who hunt and fish on their time off.

Licenses for whaling are only issued to residents in the first category. When considering the socioeconomic importance of whaling in Greenland the following factors should be taken into account:

- there are no land-based whaling operations in Greenland,
- no vessels are equipped for whaling only,
- no organized export of whale products takes place,
- market situations in or outside Greenland have little impact on the hunt, because,
- the hunt is opportunistic and subsistence motivated, not systematic and commercially motivated.

Cultural and Nutritional Attributes of Subsistence Whaling by Indigenous Peoples

In ancient Greenland whaling played a relatively minor role in the establishment of most settlements. For the development of certain larger settlements, however, whaling was important because it required participation by a greater number of people, which, in turn added stability to the larger settlements. After 1650, a marked change in the settlement pattern in Greenland occurred. This change resulted partly as a consequence of difficult whaling conditions during that period, causing significant disruptions to the economic and social structure of communities (Petersen 1974/75:176).

The importance of whaling *vis-à-vis* a person's status and role in the community is only occasionally referred to in Greenlandic literature. The song of the unfortunate harpooner, Kuukujuk, who missed his mark and therefore was displaced to the steering-oar, clearly indicates the difficulties in accepting the deprivation of the leading role (Nielsen

1980:65). That whaling had a special status is also indicated by the fact that the participants in the hunt set out in new, fine clothing (Egede 1741 [1925]:76).

Nevertheless, it is conceivable that many of the ceremonies connected with Alaskan Inuit whalers had no parallels in ancient Greenland, possibly because subsistence hunting in Greenland was more diversified. What characterizes Greenlandic whaling today is that it is an integral part of a combined hunting and fishery economy, which is directed towards utilization of as many resources as possible, thus reducing pressure on any particular species.

Whaling still constitutes an important part of nutrition and diet in Greenland (see Appendix I by Dr. P. Helms). The demand for whale meat in Greenland has probably never been fulfilled, which may account for one reason why the organized export does not take place. The unsatisfied demand in Greenland also means that whale products are more expensive than some other locally available products. Therefore, institutions with a limited budget (e.g., boarding schools and hospitals) are unable to obtain whale meat in the desired quantities. Yet, despite the fact that whale meat and *muktuk* are only available in limited quantities, the demand has not caused an increase in whale hunting.

Although no organized export from Greenland takes place, it is possible in Denmark, by special request, to purchase whale meat. The form and quantities of such transfers is, however, consistent with the definition of 'local consumption.' The bulk of Greenlandic whale products available outside Greenland is used by Greenlanders. In Denmark, whale meat is a highly appreciated contribution to social gatherings and celebrations held by Greenlandic organizations.

Management Principles Potentially Applicable to Subsistence Whaling by Indigenous Peoples

As this paper deals only with Greenland, the following considerations apply to the Greenlandic catch of fin, humpback, and minke whales. Similar management principles could also be considered in the case of the Bering Strait grey whales and the Alaska bowhead whales. Because of the different status of these whale populations, and because of variation in cultural and sociological background of the peoples involved, uniform management principles would be difficult to apply. Certain basic requirements could, however, be established for such management principles:

- Equal importance should be attached to scientifically documented conservation requirements and to the necessity of the subsistence catch to meet nutritional, social, and cultural needs.
- The principles should be established in such a way that they are likely to obtain the acceptance of the indigenous peoples involved. Otherwise, there will be a considerable risk of infractions, which the respective contracting governments will have only limited ability to effectively monitor and control.
- For this reason the principles should not be too complex.
- The IWC should not recommend measures other than quotas. For non-endangered stocks these quotas should only be 'start point' quotas. Other regulatory measures should be established on a national basis.
- For stocks other than endangered stocks there should be a certain flexibility for the national establishment of final quotas, so as to enable national decision-makers to set final quotas somewhat higher or lower than the IWC 'start point' quota. In case final quotas are set higher, increases should only be moderate from year to year so as to avoid negative effects.
- The IWC Scientific Committee should retain the possibility of forwarding new recommendations at any time, even after IWC quotas and national final quotas have been set.

Management and Regulation of Whaling

National Regulations

Final quotas for non-endangered stocks and other regulatory measures should be established on a national basis. It is considered advantageous to involve the indigenous peoples concerned as far as possible in the decision-making process. The indigenous peoples could themselves decide upon the distribution of quotas (if appropriate), closed areas, closed seasons, etc. The extent to which such decisions may be delegated to indigenous peoples will depend upon the internal constitutional arrangements in the respective contracting states.

For Greenland, the Danish Government has issued two regulations on whaling after entertaining suggestions from the Greenland Home Rule Authorities. Regulation No. 484 of November 30, 1979 for Greenland on Restrictions of Whaling Activities, sets quotas and size limits corresponding to IWC rules. It also contains the prescriptions that meat and other products from humpback whales may only be used for local consumption in Greenland, and that such products from fin and sei whales may only be used for local consumption, and for domestic animals in Greenland. Regulation No. 485 of November 30, 1979 for Greenland on Reporting Whaling Results contains provisions corresponding to IWC rules on reporting on the catch and the whales caught. Furthermore, certain supplementary rules on the catch of minke whales are in force. These rules were issued by the Greenland Provincial Council (Grønlands Landsråd), which was the predecessor of the Greenland Home Rule Authorities.

Revised Management Procedures
The report of the Technical Committee Working Group on Revised Management Procedures (Rome, May 1981) proposes the automatic phasing out of catches for which not all data requirements have been fulfilled. This proposal is unacceptable in regards to subsistence whaling because of its nutritional, social and cultural needs and because of the practical difficulties involved in gathering data sufficient for full evaluations of the exact status of the stocks.

Approaches to Setting Catch Limits for 'Stocks' Where Subsistence Whaling Occurs

Scientific Committee Advice
The present composition of the Scientific Committee of the IWC reflects the fact that the task of that committee is to 'review the current scientific and statistical information with respect to whales and whaling' (IWC Rules of Procedure, April 1981: Section J.3.). 'Scientific' in this context is interpreted as referring to biological science and such disciplines that are needed for population dynamic analyses. The kind of scientists appointed to and attending the meetings, and the way the Committee works, further reflects the fact that the Scientific Committee's main task has been to give advice on stocks, which are the object of commercial whaling. When certain questions relating to subsistence catches were recently referred to the Scientific Committee, they were dealt with by an *ad hoc* Subcommittee on Protected Species and Aboriginal Whaling, but it must be stressed that the Scientific Committee was not enlarged or supplemented by new specialists to enable it to cope adequately with these issues. The Committee itself has repeatedly expressed the view that it is not particularly suited for dealing with some of the questions relating to aboriginal subsistence whaling. The future role of the Scientific

Committee should be confined to a review of the status of stocks on the basis of available biological and statistical data, to identify data gaps, and to give advice on possible trends in future stock sizes under different regimes. These would include any subsistence take, commercial catches, or other man-made environmental disturbance (see below). The Scientific Committee should not, however, be asked to evaluate cultural, socioeconomic, or nutritional aspects of subsistence whaling.

Assessment of Cultural and Nutritional Requirements of Indigenous Peoples

As indicated in a previous section, it is considered important for the success of any management regime where subsistence whaling occurs that equal importance be attached to the documented requirements of indigenous peoples and to the advice of the Scientific Committee. It is, thus, important that the documentation and assessment of these requirements is given consideration and treatment equal to the assessment of biological and environmental aspects. This will necessitate the establishment of a mechanism whereby advice is invited from scholars from all relevant disciplines (e.g., anthropologists, demographers, nutritionists, etc.). It is, further, imperative that representatives of the indigenous peoples involved in subsistence whaling are included in this process.

A first step in this direction was taken with the holding of the Panel Meeting of Experts on Aboriginal/Subsistence Whaling in Seattle, February 1979, followed by the Technical Committee Working Group Meetings in April 1979 and February 1981. These meetings resulted in some clarification of the problems relating to the Alaskan catch of bowhead, but did not address other examples of subsistence catches adequately. It was further argued that full benefit was not derived from the presence of researchers from various fields at the first Seattle meeting (Mitchell and Reeves 1980).

It is crucial for adequate discussion of many of these issues that a mechanism be established to allow continued documentation and assessment of the requirements of indigenous peoples involved in subsistence whaling. One way of doing this could be to make the present *ad hoc* Technical Working Group on Subsistence Whaling a standing working group or subcommittee, and invite the involved nations to appoint qualified representatives to participate in its work.

R. Petersen, E. Lemke, F.O. Kapel

Reporting and Research Requirements for Subsistence Catches

General consideration of data-gathering and reporting requirements

In principle, the same data on catch levels, composition of catches and hunting effort is required for subsistence catches as for commercial whaling, but the details and the conditions for collecting them are different. Further, data on the utilization of the products and on the cultural, social, and economic importance of subsistence whaling and hunting are relevant, but discordant with rigid scientific analyses. Thus, such variables may have to be gathered and evaluated by a completely different procedure.

Statistical information on catch levels

Information on the number of marine mammals caught in Greenland have been collected since the last decades of the 19th century through the so-called 'Greenlanders' Lists of Game' (cf. Kapel 1978a:217, 1979:198). The system functions by appointing a person in each settlement to keep record of the hunting results of all individuals participating in the hunt, by species and half month, and to deliver the reporting forms to the local authorities who forward them to the Ministry for Greenland at the end of the year. Originally, the lists only contained information on the number of seals and caribou taken, but later on data on small cetaceans were added. Since 1950 catches of minke whales and larger whales were also included. The 'Lists of Game' do not give complete coverage in all years for all regions, but they represent a unique data base for evaluating the importance of hunting for a period of more than one hundred years.

When some Greenlandic fishing vessels were equipped with harpoon cannons around 1950, their owners were asked to provide annual reports on the whales taken (cf. Kapel 1978a, 1979: 'Whaling Reports'). Good reports were received for a number of years, but in recent years the coverage has been poor. In 1980 a new 'Government Notice' was issued in an attempt to reconstruct and improve this reporting system.

In summary, previous and current arrangements for collecting statistical data on whale catches in Greenland have been based on information given by the hunters themselves or by another person in the local community. The latter approach has up to now proved more successful than the former. Both systems are based on principles of deliberate cooperation by the local peoples, not on rigid proclamations or penalty clauses for lack of compliance.

The possibility of appointing official inspectors in order to improve the existing reporting systems has been suggested, but considering the variable and scattered nature of the operations, this will be an extremely expensive and complicated solution. It is considered more realistic and profitable to continue the gradual improvement of the existing systems, possibly combined with an evaluation process involving local authorities and independent experts.

Information on Composition of Catches and Hunting Effort

The 'Lists of Game' do not include information on the sex or the length of the whale, on methods and equipment used, or indications of the time and effort involved in catching. Some data of this kind have been recorded in the 'Whaling Reports,' but as mentioned above they only cover a fraction of the catches. The new 'Government Notice' includes requests for some of this information.

It is, however, questionable how reliable these data are or are likely to be in future. In evaluating this kind of information obtained from the hunters it must be taken into account that the vessels are small, engaged in activities other than whaling, and with only enough crew to run the boat and conduct fishing and hunting operations. There is little extra capacity or room for maintaining extensive reporting procedures or personnel on board the vessel. Any reporting is likely to take place at the end of a trip or later. This implies that there will be a limit to the amount of information expected from these operations on a routine basis.

In summary, and for the time being, it is unrealistic to put much confidence in information from the kind of data referred to in this section through a rigid, routine reporting system. A better approach may be to include collecting such data as part of a research program.

Information on Utilization of Products and on Socioeconomic Values of Hunting

The last statement is even more obvious in respect to collecting information on the utilization of whale products, and the present importance of hunting for the Greenland community. It may be possible to persuade hunters to give some estimates of the utilization of the whale products, e.g., the approximate amount of meat and blubber used for human consumption and for dog feed. In the exceptional case, where whale products are sold to a freezing plant for distribution in other regions of Greenland, it will be possible to get more exact data for products not consumed locally, but equivalent data on the part of the catch used domestically will be impossible to get on a routine basis.

The question of the cultural importance of whaling, its economic and social value for the individual hunter, the local settlement and the entire Greenland community, is so complicated that routine collection of data (e.g., the market value of a given whale) is meaningless or even misleading. Thus, such efforts must be closely linked to proper research projects in an attempt to place the information in a coherent and appropriate context.

Subsistence Whaling in Greenland

General Considerations of Research and Monitoring Requirements

It is obvious that only a minor part of the information required for proper management can be obtained from routine data-gathering by hunters or other local residents, and that especially designed research is necessary to collect additional data. The current reporting systems may, with recently introduced improvements, give sufficiently exact data on the number of whales of each species taken per vessel/settlement and the approximate time (month) for the catch. They may also give some data on location of catches, sex and length of the animals, presence of a foetus, number of whales struck and lost, and a rough indication of the utilization of products. For most, if not all, of this information it is important for qualified personnel to evaluate the quality of the data through fieldwork. Such research will also include the collection of additional material, e.g., biological species, data on hunting methods, hunting effort, quantitative data on the utilization of products, and information related to the question of the socioeconomic value of hunting.

In designing appropriate research programs a number of logistic problems arise. A major obstacle is that whale hunting in Greenland is mainly a secondary activity carried out in an opportunistic way in conjunction with other hunting and fishing activities. It is almost impossible to predict when and where a significant number of whales will be caught, and by whom. In the case of minke whaling, for instance, more than one hundred fishing vessels are equipped with harpoon cannons, but only half of them catch one or more minke whales in a given season (cf.

Kapel 1978a). A vessel may take one or two whales early in the season, but next year it may not catch a single whale until late in the autumn. Catches are scattered along a fairly long coastline. These facts suggest that it is virtually impossible to carry out research which covers more than a small fraction of the total catch.

The collection of some of the data mentioned above depends on either being aboard the catching vessel, or on being able to follow it closely in another boat. The first condition will depend on the willingness of the owner of the vessel to take a passenger, and will in many cases be impossible to arrange because of the small size of the vessel. The second situation may interfere adversely with hunting and potentially dangerous situations may arise. A third alternative, namely to wait ashore until the whale is landed for treatment may offer better possibilities for collecting important material and data. It is, however, often undecided which locality is going to be used for flensing, until the whale has been caught. A research team will, therefore, have to establish close cooperation and radio contact with the catching vessels, and have access to a fast boat in order to be able to reach the flensing locality in time.

Among other logistic difficulties, it may be mentioned that transportation between settlements in Greenland is often slow, unreliable and expensive, and that accommodation facilities are not available in most settlements. Finally, it should be stressed that in order to communicate with the hunters and other local residents, at least one member of a research team must have some ability in understanding and speaking Greenlandic.

Independent Observation of Subsistence Catches

Independent observation can be defined as observation by persons other than those directly involved in the hunting or dependent on the products. It may be carried out by researchers, national inspectors, or international observers. The considerations put forward in the preceding section should apply to all independent observers.

Possible Impact on Traditional Hunting

The mere presence of outside observers could create a feeling by the local residents of unnecessary interference in their internal affairs, particularly in the smaller settlements. It is extremely important that the local population understands and accepts the purpose and possible consequences of admitting external observers. It is equally important that the observers are aware of this problem, and are able to participate in an exchange of points of view during their stay. The more the observers want to be present in the actual hunting situation, the more their presence is likely to have an impact on the hunting. For this reason the number of observers should be kept to a minimum. In order to get maximum benefit from an observer team, the team should be responsible for collecting as much information as possible, although this may appear to conflict with the principle of distinguishing between researchers and observers/inspectors.

Logistic Arrangements

It is obvious that it will not be possible to observe and research all whale hunting in Greenland, unless a major commitment in time and money is devoted to the task. It is unrealistic for one observer team to examine more than the catch of 10-15 minke whales or larger species of whales within a two month period, which is less than 10% of the total catch.

Three different approaches may be considered for collecting data on the catch of whales in Greenland. One would be to work from one or more of the larger fishery settlements in southwest Greenland: Paamiut (Frederikshab), Nuuk (Godthab), or Maniitsoq (Sukkertoppen) districts. Some of the vessels catching minke, and occasionally fin and humpback, are large enough to accommodate one or two extra passengers during the hunting trips. Such an arrangement might offer opportunities for researchers to study the relative effort put into hunting and fishing by these vessels. Qualified interpreters may be found more easily in this region than elsewhere in Greenland. For a suggested period of two months (summer or autumn), half the time will probably be waiting time with little related work to do. Accommodations ashore probably could be arranged easily.

A second approach would be to settle down at a locality where whales are brought in for flensing at a regular basis. This would offer excellent opportunities for collecting samples and data on whales landed by different vessels, but the chance of participation in the hunt itself would be minor, and there is a risk of waiting in vain for a longer period. (An attempt to collect material in this way was made by Greenland Fisheries Investigations in 1978 at Qeqertarssuaq (Godhavn), but during a period of six weeks only seven whales were examined.) The arrangement may also offer good possibilities of studying other hunting or fishing activities. In some localities (e.g., Qeqertarssuaq) arrangements for modest accommodation could be made if notice is given well in advance. In other places this may not be the case, and the research team might have to erect its own camp outside the settlement.

The third approach would be to erect a camp somewhere in the western part of the Umanak district. Here, 4-5 small vessels are taking 15-30 minke whales per summer. Whales are then flensed at a number of localities with sloping beaches, often far from the settlements. The research team must have one (or better two) small, fast motor boats at its disposal as well as a system of radio communication with the hunting vessels and with a contact person in Umanak. It may even be possible to arrange *ad hoc* participation in the hunting aboard one of the vessels. This approach will present good possibilities of studying other hunting activities (seal and seabird) as well as the local use and distribution of products. As indicated, accommodation must be arranged in the form of camping.

Whether one or a combination of approaches is chosen, the establishment of an observer/research program will have to overcome a number of logistic difficulties. In the end, the coverage will be small and the cost will be considerable compared with the results.

Inclusion of Subsistence Whaling in an Approved IWC Observer Scheme

The prime purpose of the existing IWC Observer Scheme is to supplement national inspection with an international reporting system in order to ensure that regulations are followed in practice. The observer scheme is not considered an arrangement for conducting additional research.

In the case of subsistence catches the main purpose of international inspection might be to document that catches are, in fact, taken for subsistence purposes. However, owing to logistic complications related to working in small communities, and the general lack of information on whaling in Greenland, it is considered appropriate that observers under these circumstances also participate in the collecting of all material and data required for evaluating the various aspects of the hunt, its importance, and consequences.

These considerations lead to the conclusion that the qualifications of persons appointed to observe subsistence catches should most likely be different from those serving in commercial operations. In Greenland, the qualifications of external observers should include some background knowledge of Inuit culture and language. In considering the possible extension of the IWC Observer Scheme to include subsistence catches, it must be taken into account that such an extension will have financial consequences for local authorities, who will be forced to establish bases of operation in setting up relevant research projects and suitable logistic facilities.

It is implied in the above sections that any effort to improve data gathering, conduct research, and introduce international observers should be well-coordinated, with the intention of providing the optimum background information on all aspects of the subsistence hunt.

Examination of Killing Techniques

IWC workshop on humane killing techniques for whales

The report of the Workshop on Humane Killing Techniques for Whales, held in Cambridge, November 1980, contains a number of recommendations, some of which may also be pertinent to subsistence catches (e.g., research into the use of high velocity projectiles). One of the recommendations (# 7) directly refers the question of possible improvements in such techniques to the *ad hoc* Working Group on (Management Principles for) Subsistence Whaling. In the case of subsistence catches in Greenland, it should be stated that few possibilities of conducting research in new methods or techniques are anticipated, but results from such research carried out elsewhere may be highly relevant to whale hunting in Greenland. In considering the possibilities of introducing new methods for the small scale hunting of whales in Greenland it must be born in mind that the practical conditions and the economical background for changing the traditional methods are quite different from those of commercial whaling.

Other considerations

Although methods currently used for killing whales in Greenland are known in a general way, and have been referred to above, there is an apparent lack of exact data on their practical implications. In respect to the conditions for gathering these data in Greenland the same reservations must be put forward as mentioned in the previous section. It is obvious that the collection of data on killing methods and techniques in Greenland must be coordinated with gathering of other kinds of data.

Bibliography

Amdrup, G.C. *et al.* (editor). 1921. Grønland i Tohundredåret for Hans Egedes Landing, I-II. *Meddr. om Grønl.* 60-61:741 + 795 pp.

Egede, H. 1741 .1925. Det gamle Grønlands nye Perlustration eller Naturel-Historie .... (ed. L. Bobé). *Meddr. om Grønl.* 54: 305-404.

Egede, P. 1741-1770. 1939. Continuation af Hans Egedes Relationer.... etc. (ed. H. Ostermann). *Meddr. om Grønl.* 120.

Fabricius, O. 1818a. 1962. Zoologiske Bidrag, Andet Bidrag: Om Stubhvalen, Balaena Boops. *Kgl. Da. Vidn. Selsk. Skr.* 6 (1): 63-83. (ed. E. Holtved, Otto Fabricius'ethnographical works. *Meddr .om Grønl.*, 140 (2), 140 pp.)

Fabricius, O. 1818b. 1962. Nöiagtig Beskrivelse over Grønlændernes Land dyr-, Fugle- og Fiskefangst med dertil hørende Redskaber. *Kgl. Da. Vidn. Selsk. Skr.* 6(2): 231-273. (Meddr om Grønl., 140 (2): 140 pp.)

Glahn, H.C. 1921. Missionær H. C. Glahn's Dagbøger for Aarene 1763-64, 1766-67 og 1767-68. H. Ostermann (ed.). *Grønl. Selsk. Skr.* 4, 247 pp.

Hansen, R. (ed.). 1971 (1922). Grønlandske fangere fortæller. *Nord. Land. Bogforl.*, 206 pp.

IWC (International Whaling Commission). 1978. Denmark. Progress report on whale research, 1976-77. *Rep. Int. Whal. Comm.* 28:101.

IWC. 1979. Denmark. Progress report on whale research, 1977-78. *Rep. Int. Whal. Comm.* 29:115.

IWC. 1981. Denmark. Progress report on whale research, 1978-79 and 1979-80. *Rep. Int. Whal. Comm. 31.*

Kapel, F.O. 1975c. Preliminary notes on the occurrence and exploitation of smaller cetacea in Greenland. *J. Fish. Res. Bd. Can.* 32 (7):1079-1082.

Kapel, F.O. 1977b. Catch of belugas, narwhals and harbour porpoise in Greenland, 1954-75, by year, month and region. *Rep. Int. Whal. Comm.* 27: 507-520.

Kapel, F.O. 1977c. Catch statistics for minke whales, West Greenland, 1954-74. *Rep. Int. Whal. Comm.* 27: 456-459.

Kapel, F.O. 1978. Catch of minke whales by fishing vessels in West Greenland. *Rep. Int. Whal. Comm.* 28: 217-226.

Kapel, F.O. 1979. Exploitation of large whales in West Greenland in the 20th century. *Rep. Int. Whal. Comm.* 29: 197-214.

Kapel, F.O. 1980. Sex ratio and seasonal distribution of catches of minke whales in West Greenland. *Rep. Int. Whal. Comm.* 30: 195-199.

Kapel, F.O. and R. Petersen, R. 1979. Subsistence Hunting: The Greenland Case. Paper No. 4 presented to the Panel Meeting of Experts on Aboriginal/Subsistence Whaling, Seattle, USA, Feb. 1979.

Kleivan, H. 1964. Acculturation, ecology and humane choice. *Folk* 6 (2).

Mitchell, E.D. and R.R. Reeves. 1980. Overview of aboriginal and subsistence whale fisheries, and analysis of the Alaskan bowhead problem. *IWC Paper* SC/32/PS 22.

Nielsen, K. J. (ed.). 1980. *Erinarsuutit.* Nuuk.

Petersen, R. 1974/75. Some considerations concerning the Greenland longhouse. *Folk* 16-17.

Rink, H.J. 1852. Grønland geographisk og statistisk beskrevet. I. Det nordre Inspektorat, Første Del. 202 pp.

Rink, H.J. 1852. Grønland geographisk og statistisk beskrevet. II. *Det søndre Inspektorat.* 402 pp.

Rink, H.J. 1877. (1974). *Danish Greenland, its people and products.* (new edition) C. Hurst & Co., London and A. Busch, Copenhagen.

Rosing, O. 1976. *Taserᵃlik.* Nuuk.

APPENDIX I

Nutritional Needs Relating to Aboriginal Subsistence Whaling Among the Inuit in Greenland

Peder Helms, M.D.
Institute of Hygiene
University of Aarhus, Denmark
1983

The Value of Whale Meat in the Greenland Diet

Since precolonial times, when Greenlanders lived almost exclusively from animal products, meat has formed a significant part of the diet. According to revised catch records (The Ministry of Greenland: abstract of catch records etc. 1977), the annual meat consumption of the resident Greenland population in terms of kilograms per capita a year is 107 kg (Table 1).

To this may be added meat and meat products imported to Greenland from Denmark, which in 1977 reached a value of Dkr 38,551,000. If the price of imported meat is estimated at Dkr 30/kg, 25 kg are consumed per capita per year, if the figure is based on the resident as well as the non-resident population. The average meat consumption per capita in Denmark in 1977 was 76 kilograms. If the same figure is applied to Danes living in Greenland in 1977 and the Danish meat consumption extracted from the total import, the result will be 14 kg of imported meat

Table 1.
Available Meat for Consumption in Greenland

Seal meat	Whale meat	Reindeer and bear	Sheep*	**Total**
61.6	29.4	8.6	7.4	107 kg per capita/year

*estimated value of meat from 8,750 slaughtered sheep and lambs

per resident Greenlander. This figure may be compared with a calculation of the 1977 budgets of Greenland families of which 7.5% of all food expenses on meat correspond to 13 kg/capita of meat purchased. If these 13 kg are added to the above 107 kg, the annual consumption per Greenlander born in Greenland is 120 kg of meat per year. Whales supply 1/4 of the meat consumed in Greenland (Table 2). Yet, there are wide differences in meat consumption in various parts of Greenland (Tables 2 and 3).

The Nutritional Value of Whale Meat

Table 4 shows the content of energy, fat, and protein in one kilogram of whale meat compared with one kilogram of seal meat and one kilogram of 'Danish meat.' The table also compares the three types of meat as to amount of energy (= satiate value, here 10,000 kJ) supplied by one kilogram of 'Danish meat,' 1.2 kg. of seal meat, and 1.5 kg. of whale meat. The differences in weight are mostly determined by the fat content, which in whale meat supplies 35% of the energy, as against 48% in seal meat. 'Danish meat' has a fat content of 70% of the energy. According

Table 2.
Seal and Whale Meat Consumption in Various Parts of Greenland

Production of	seal meat/kg/capita/year	whale meat/kg/capita/year
South western Greenland	9	25
North western Greenland	111	36
Northern Greenland	362	66
Eastern Greenland	115	3

Table 3.
Average Catch of Whales and Small Cetaceans in Greenland, 1975-79, and Amount of Products for Consumption from this Catch by Species and Region.

	N.W. Greenl.	C.W. Greenl.	S.W. Greenl.	S. Greenl.	E. Greenl.	Total
Average catch (range)						
Minke whale	41.4 (30-56)	64.6 (46-89)	80.0 (58-95)	39.6 (17-71)	0.6 (0-1)	226.2 (180-286)
Fin whale	0	2.8 (0-8)	5.4 (2-7)	0.2 (0-1)	0	8.4 (4-14)
Humpback whale	0.2 (0-1)	0.4 (0-2)	13.8 (9-23)	0	0	14.4 (9-23)
Pilot whale	0	0.6 (0-2)	85.6 (47-134)	2.6 (0-10)	0	88.8 (50-136)
Narwhal	272.4 (199-349)	103.2 (44-262)	3.8 (0-9)	0.2 (0-1)	15.6 (3-24)	395.2 (278-615)
White whale	232.4 (139-339)	486.2 (268-953)	114.4 (65-167)	0.2 (0-1)	0.8 (0-2)	834.0 (656-1213)
Harbour porpoise	1.2 (0-3)	160.4 (69-222)	775.4 (711-865)	8.4 (0-21)	0	945.4 (789-1075)
Products, Kg.						
Minke whale	82,800	129,200	160,000	79,200	1,200	452,400
Fin whale	0	28,000	54,000	2,000	0	84,000
Humpback whale	1,600	3,200	110,400	0	0	115,200
Pilot whale	0	240	34,240	1,040	0	35,520
Narwhal	61,290	23,220	855	45	3,510	88,920
White whale	46,480	97,240	22,880	40	160	166,800
Harbour porpoise	24	3,208	15,508	168	0	18,908
Total =	192,194	284,308	397,883	82,493	4,870	961,748
Products per capita						
Minke whale	14.99	11.08	8.14	9.54	0.39	9.38
Fin whale	-	2.40	2.75	0.24	-	1.74
Humpback whale	0.29	0.27	5.62	-	-	2.39
Pilot whale	-	0.02	1.74	0.13	-	0.74
Narwhal	11.09	1.99	0.04	0.01	1.14	1.84
White whale	8.41	8.34	1.16	0.000	0.05	3.46
Harbour porpoise	0.00	0.28	0.79	0.02	-	0.39
Total =	34.79	24.38	20.25	9.94	1.58	19.95

Table 4.
Comparison of Whale-Meat, Seal-Meat, and Danish Meat

		Energy	Protein	Fat	Fatty Acids saturated	Fatty Acids polyunsaturated	P:S ratio
Content per Kilogram		kJ	g	g	g	g	
Whale meat		6,530	250	60	12	8	0.67
Seal meat		8,050	250	100	16	22	1.38
Danish meat [Bacon 70% Beef 18.5% Chicken 11.5%]		10,000	170	185	75	20	0.26
Content per 10,000 kJ							
Whale meat	(1.5 kg)	10,000	385	92	18	12	0.67
Seal meat	(1.2 kg)	10,000	310	125	20	27	1.38
Danish meat	(1.0 kg)	10,000	170	185	75	20	0.26

to Scandinavian standards, a low-fat diet is preferable corresponding to a maximum of 35% of the energy in the total diet. If this evaluation is applied to the three types of meat, whale meat is obviously the best choice.

Not only the quantity of the fat, but also its quality is essential in determining the health value of the food. A high ratio of polyunsaturated to saturated fatty acids is significant in the prevention of cardio vascular diseases. These are still far rarer in Greenland than in Denmark and other parts of Europe.

Investigations made by Bang and Dyerberg (1981) have shown an extremely favourable ratio of polyunsaturated to saturated fatty acids in the fats in the Greenland hunter diet. This ratio for the three types of meat appearing in the table may be calculated at 0.67 for whale meat, 1.38 for seal meat, and 0.26 for 'Danish meat.' The 'Nordiske * Naeringsstofanbefalinger' (recommended intake of nutrients in Scandinavia) aims at a ratio of 0.50 in the Danish average diet.

If whale meat were no longer included in the Greenland diet, but were replaced by meat from Denmark, the diet would be more fatty and have a lower ratio of polyunsaturated to saturated fatty acids. Both of these modifications in the traditional Greenlandic diet would imply introduction of risks for atherosclerotic and cardiovascular diseases in a population that until now has had a low prevalence of such diseases.

List of References

Ackman, R.G. and S.N. Hooper. 1974. Long-chain monoethylenic and other fatty acids in heart, liver, and blubber lipids of two harbour seals (*Phoca vitulina*) and one grey seal (*Halichoerus grypus*). *J. Fish. Res. Board Can.* 31: 333-341.
Ackman, R.G., C.A. Eaton, and P.M. Jangaard. 1965. Lipids of the fin whale (*Balaenoptera physalus*) from North Atlantic waters. (Fatty acid compositon of whole blubber and blubber sections). *Canadian Journal of Biochemistry* 43: 1513.
Ackman, R.G., C.A. Eaton, and P.M. Jangaard. 1965. Lipids of the fin whale (*Balaenoptera physalus*) from North Atlantic waters. (Fatty acid composition of the liver lipids and gas liquid chromatographic evidence for the occurrence of 5, 8, 11, 14-nonadecatetraenoic acid.). *Canadian Journal of Biochemistry* 43: 1521.
Ackman, R.G., S. Epstein, and C.A. Eaton. 1971. Differences in the fatty acid compositions of blubber fats from Northwestern Atlantic finwhales (*Balaenoptera physalus*) and harp seals (*Pagophilus groenlandica*). *Comp. Biochim Physiol.* 40B: 683-697.
Anon. 1983. Leading article: Eskimo Diets and Diseases. *Lancet* 1: 1139-1141.
Bang, H.O. and J. Dyerberg. 1975. Blodets fedtindhold og kostens sammensætning hos en vestgrønlandsk befolkningsgruppe. *Ugeskr. Læger* 137: 1641-1646.
Bang, H.O. and Dyerberg. 1981. The Lipid Metabolism in Greenlanders. *Meddelelser om Grønland, Man & Society* 2-1981: 3-18.
Bang, H.O., J. Dyerberg, and N. Hjørne. 1976. The composition of food consumed by Greenland Eskimos. *Acta Med. Scand.* 200: 69-73.
Bang, H.O., J. Dyerberg, and H.M. Sinclair. 1980. The composition of the Eskimo food in northwestern Greenland. *Arn. Journ. Clin. Nutr.* 33: 2657-2661.
Dyerberg, J. and H.O. Bang. 1978. Dietary fat and thrombosis. *Lancet* 1978 I: 152.
Dyerberg, J. and H.O. Bang. 1982. Factors influencing morbidity of acute myocardial infarction in Greenlanders. *Circumpolar Health*, 5th International Symposium, Nordic Council for Arctic Medical Studies, Rep. Ser. 33:300-303.

Dyerberg, J., H.O. Bang, E. Stoffersen, S. Moncada, and J.R. Vane. 1978. Eicosapentaenoic acid and prevention of thrombosis and atherosclerosis? *Lancet* 1978 II:117-119.

Goddard, V.R. and L. Goodall. 1959. *Fatty acids in animal and plant products.* Washington, DC.

Helms, P. 1980. *Kostvurderingstabeller*, Akademisk Forlag. København.

Helms, P. 1981. Kostundersøgelse i Angmagssalik. *Forskning/tusaut i Grønland* 1-2:10-14.

Helms, P. 1981 Angmassalingme nerissanik misigssuineq. *Forskning/tusaut i Grønland* 1-2:14-18.

Helms, P. 1982. Changes in disease and food patterns in Angmagssalik, 1949-1979. *Circumpolar Health*, 5th International Symposium, Nordic Council for Arctic Medical Studies, Rep. Ser. 33:43-251.

Ministeriet for Grønland: Samendrag af fangstlister m.v. 1976, 1977, 1978.

Ministeriet for Grønland: Arsberetning: Grønland 1981, 12, 1982.

Statistisk Arbog, Danmarks Statistik, 1976 to 1982.

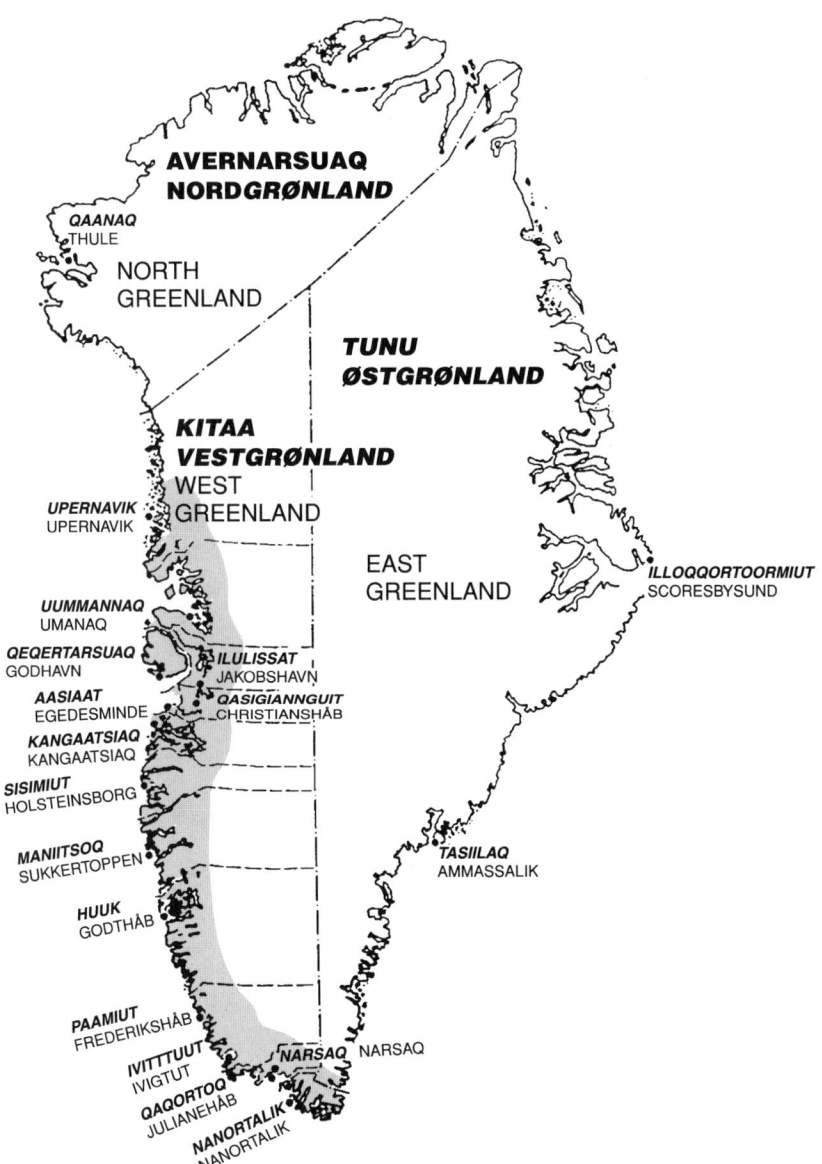

Community-based Whaling in West Greenland.

Chapter 3

The Greenland Aboriginal Whale Hunt[1]

Report to the Standing Committee on Aboriginal/Subsistence Whaling of the International Whaling Commission

P. Helms, M.D.
Institute of Hygiene, University of Aarhus

O. Hertz
Institute of Eskimology, University of Copenhagen

F.O. Kapel
Greenland Fisheries Investigation, Copenhagen
1984

---

1. Editoral note: Not quoted *in extenso*; original paper has been extensively modified and reorganized.

P. Helms, O. Hertz, F.O. Kapel

Introduction

The taking of large baleen whales as well as smaller toothed whales has for centuries been part of the hunting pattern which was the basis for existence in Greenland. The earliest trace of the use of whale products goes back to around 2000 BC (Melgaard 1983). This hunting pattern also included exploitation of other marine organisms (primarily seals) and some terrestrial animals. It is therefore necessary to consider whaling in Greenland in relation to the exploitation of other living resources.

Originally, all hunting was carried out by means of 'primitive' tools, although some of these were highly adapted to various types of hunting using available raw materials (Fabricius 1810, 1818; Holtved 1962). As a result of contact with European peoples during the 18th and 19th centuries, a modification of the original hunting methods took place (e.g., the introduction of firearms). In general, however, methods remained unchanged up to the beginning of this century. The recent development of a modern fishery in southwestern Greenland has had some influence on the hunting districts. But on the whole, hunting has kept its traditional character as a technologically simple, locally oriented occupation.

Hunting in Greenland has always, with few exceptions, been carried out by the indigenous people of Inuit origin, although for several hundred years some mixing with European peoples has taken place (the common term for the indigenous people is 'Greenlanders', or the more recently re-introduced, 'Inuit'). The purpose of hunting was, of course, to provide the basis for existence. Because of seasonal and long-term fluctuations in the availability of the prey animals, it was necessary to have means of smoothing out the differences between periods of high and a low hunting yields. This was done partly by food conservation and storage, partly by the exploitation of a diversity of species. It is important to stress that domestic hunting can normally be adapted to seasonal differences without expensive technological adjustments and that hunting in Greenland still does not support industries based exclusively on the commercial exploitation of animals.

For most of the Greenlandic population, fishing is today more important economically than hunting, and the export of fishing products plays a different role in the economy than that of hunting products. Fishing requires a large degree of modernization and investment in industrial plants and efficient vessels. The demands of fish processing industries tend to create stable, or even increasing exploitation of fish resources. Greenlandic fishermen and commercial fishing are vulnerable to changes in external market conditions, very much dependent on international quotas and regulations, and sensitive to small changes in the environment. The protection of renewable resources is therefore an important objective for fisheries policy in Greenland.

Present-day hunting in Greenland is, of course, also influenced by technical innovations and economic considerations. Technical improvements and needs in hunting have made a monetary income necessary even in the hunting districts. Also, the traditional system of mutual exchange of hunting products does not work in all cases; some households must reciprocate with money as they have no hunting products to give in return. Although hunting is also an important supplementary occupation in the fishing districts, there is a permanent demand for hunting products in these areas. Food products from the hunt are still valued highly in all parts of Greenland, and constitute an important contribution of hunting in Greenland. In order to evaluate the importance of present day whaling and hunting activities in Greenland, it is necessary to consider many factors: the current status of living resources, occupation and demographic patterns, social and economic needs and conditions, and nutritional needs.

Subsistence Needs

Human Population: Occupational Regions and the Distribution of Whaling

Greenland can be divided into several regions, which are rather different from each other in current and potential occupation, standard of living, economy, and population densities (Figure 1). The case is complicated, however, because the differences are not only to be found between the regions, but within regions where conditions can be very different between towns and settlements. Except from the southernmost part of Greenland where some families are sheep-farmers, one place in central west Greenland where caribou are reared, and another place where a lead and zinc mine is located, all other primary occupations are dependent on the biological resources of the sea. A survey of these regions has previously been given in various publications (e.g., Kapel 1975), but is repeated below with special reference to the relative importance of whale hunting within each region. The regions where whaling occurs are listed in Table 1.

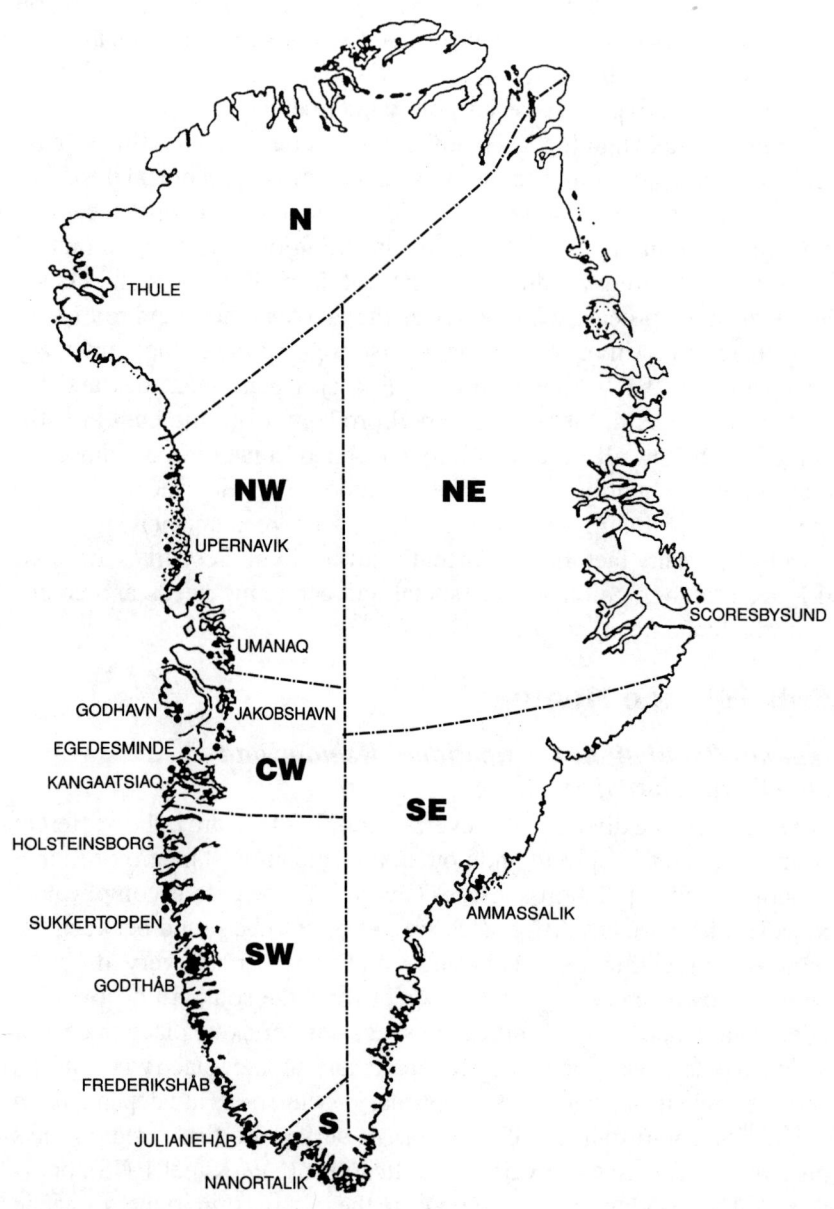

Figure 1. The main districts of Greenland.

Table 1.
Inhabited Places Where Whales Have Been Caught 1977-1981

Region S:	Small whales	Minke, fin or humpback whales
Qaqortoq	+	+
Saarloq	+	+
Eqalugaarsuit	+	+
Qagssimiut	+	+
(Sheep farm)	+	
Nanortalik	+	+
Aappilattoq	+	+
Narsaq kujalleq	+	+
Tasiusaq	+	+
Ammassivik	+	+
Alluitsoq	+	
Alluitsup Paa	+	+
Narsaq	+	+
Igaliku	+	
Region SW:		
Paamiut	+	+
Arsuk	+	+
Narsalik	+	
Avigaat	+	+
Nuuk	+	+
Qeqertarsuatsiaat	+	+
Maniitsoq	+	+
Atammik	+	+
Napasoq	+	+
Kangaamiut	+	+
Sisimiut	+	+
Sarfanguaq	+	
Region N:		
Qaanaaq	+	
Savissivik	+	
Qeqertat	+	
Siorapaluk	+	
Region NE:		
Ittoqqortoormiit	+	+
Illukasiit	+	
Region CW:		
Kangaatsiaq	+	+
Attu	+	
Iginniarfik	+	

Table 1. (continued)
Inhabited Places Where Whales Have Been Caught 1977-1981

Ikerasaarsuk	+	
Niaqornaarsuk	+	+
Aasiaat	+	+
Kitsissuarsuit	+	+
Akunnaq	+	+
Qasigiannguit	+	+
Ilulissat	+	+
Saqqaq	+	+
Qeqertarsuaq	+	+
Kangerluk	+	+
Region NW:		
Uummannaq	+	+
Niaqornat	+	+
Qaarsut	+	+
Ikerasaq	+	+
Saattut	+	+
Ukkusisat	+	+
Illorsuit	+	+
Nuugaatsiaq	+	+
Upv. kujalleq	+	+
Kangersuatsiaq	+	+
Aappilatoq	+	
Tussaaq	+	
Naajaat	+	
Innaarsuit	+	+
Tasiusaq	+	+
Nutaarmiut/Iker.	+	
Nuussuaq	+	
Kullorsuaq	+	
Region SE:		
Tasiilaqq	+	
Isortoq	+	
Tinteqilaaq	+	
Ikatteq	+	
Kulusuk	+	
Kuummiut	+	
Sermiligaaq	+	

(There is no information from 16 places).
(From Anon. 1977-1981. Sammendrag af Grønlands fangstlister m.v.)

North Greenland (N; Thule district):
Hunting is the only basis for existence in this region. The ringed seal is the most important species, but the bearded seal and walrus are also of great importance. Harp and hooded seals are only caught in small numbers. The catch of narwhal, and to some degree also beluga, is extremely important in the summer months. The polar bear is more important here than in any other part of Greenland. The catch of sea birds, and in particular the little auk, plays an important role locally, while the catch of the Arctic fox is often significant.

North East Greenland (NE; Scoresbysund district):
Hunting is the dominant occupation in this district (see Larsen, this volume). The ringed seal is by far the most important species. The polar bear and muskox are of some importance, in addition to narwhal and beluga.

South East Greenland (SE; Angmagssalik district):
Hunting remains the primary occupation here. The ringed seal is the most important species but hooded seals are also caught in significant numbers during the summer. In addition, some bearded seals, harp seals, polar bears, Arctic foxes, narwhal, and beluga are caught. There is a small fishery in one settlement.

North West Greenland (NW; Upernavik and Umanaq districts):
Hunting is the dominant occupation, although in the town of Umanak (Uummannak) and at some settlements commercial fishing for Greenland halibut, wolfish (Anarhichas), and Greenland shark is increasingly important economically. The ringed seal is the most important species during winter and spring, whereas the harp and hooded seals are of importance in summer and autumn. The catch of beluga and narwhal plays a significant role in the autumn, but less so in the spring. The minke whale is an important meat source during the summer, especially in the Umanaq district, and the same is true of birds, especially the thick-billed murre and eiders.

Central West Greenland (CW; Disko Bay and adjacent areas):
Fishing, especially for deepwater prawn and Greenland halibut, is the most important occupation in the towns of this region. However, hunting is an important supplement in the larger centres of the region and is of basic importance for subsistence in the smaller settlements, especially in winter. Ringed and harp seals are the most important species, although

hooded seals and walrus play a minor role. Beluga and narwhal are of great importance in winter and spring, particularly in years of 'savssat' (the mass-occurrence of holes in the ice). The catch of minke whales and occasionally fin whales, is locally of great importance in the summer, when bird hunting is also an important supplement to the diet.

South West Greenland (SW; Holsteinsborg to Frederikshåb district):
The most important occupation is fishing, especially for cod and deepwater prawn, but for other species as well. The fishery is carried out as a coastal fishery by small vessels, as well as an offshore trawler fishery on the banks of the Davis Strait. The catch of minke whales, and occasionally humpback and fin whales, is a significant supplement to the nutrition of the local population. The catch of smaller cetaceans (harbour porpoise, beluga, etc.) and seals (ringed, harp and hooded) plays a similar, but less important role. Bird hunting contributes much to the local nutrition, especially in the autumn. Caribou hunting is very important in winter and late summer.

South Greenland (S; Narssaq, Julianehåb, and Nanortalik districts)
At the end of the fiords, sheep farming is the important occupation for a scattered population. In the archipelago at the mouths of the fiord system, fishing is the dominant occupation, supplemented in the spring and early summer by a significant catch of seals, especially hooded seal. A small number of polar bears is regularly taken. Bird hunting is of local importance, while catches of whales and smaller cetaceans occur sporadically.

Number of Communities Engaged in Whaling and Dependence on Whaling

In all the above-mentioned regions of Greenland, whales are hunted. Whaling activities are sometimes planned and organized, sometimes not. Catches of beluga and narwhal from the ice edge or from shore are planned. But in open waters, the hunts often begin when, by chance, a whale is observed by a hunter or fisherman. A call is made to other boats for assistance by 'walkie-talkie', with those nearby abandoning their activities to be able to participate in the hunt. Bigger boats with harpoon guns can be used to carry out the hunt alone, but sometimes in cooperation with owners of smaller boats. Another procedure involves hunting the whale by small boats with outboards and then having it towed to the settlement by a bigger fishing boat. All hunters and fishermen can take part in the hunt, and everyone in the local community may, depending on the size of the population, be involved.

Over the last five years, as indicated by the available information, whaling has taken place in 71 different towns or settlements in Greenland (Table 1). The total number of inhabited places is today 120. The exact number of people presently engaged in whaling cannot be determined.

The Importance of Whaling

As hunters and fishermen, the local population in Greenland does not and cannot obtain its food by manipulating the productivity of their environment. In order to survive in the Arctic in this way, it is necessary to procure resources when they are near the local community. With the present technology it is only the bigger fishing vessels that can take resources further away from communities.

Arctic biological resources are not stable year round, or from year to year. The variation in availability is due not only to movements of the animals, but also to changing weather conditions, especially ice and wind. Seasonal and long-term fluctuations in catches are facts which must be taken into account by hunters and fishermen, and which may differ significantly from one area to another. Even between neighbouring localities, the seasonal availability of some prey animals may differ considerably.

Even if information about local ecological strategies is lacking, it is possible generally to say that hunters and small-scale fishermen seek to maintain as broad an ecological base as possible by using several species of animals. In this way, they can accommodate changes in the abundance of one species by using another more or less intensively. As examples of this multispecies exploitation strategy, it can be noted that, in the southern part of the NW region and in some small settlements of the CW region, between 50 and 70 species of animals are used (Hertz 1975, 1977a, 1977b). However, multispecies exploitation is only possible to a certain degree, because some species are more dominant in the ecosystem, or because certain species are present at a time of the year when no others are available. This is the case with the minke whales caught in the Umanaq-district (NW region). Here the whales are taken at a time of the year when there is a scarcity of meat from seals and birds. In other parts of Greenland whaling is, in similar ways, part of a fixed seasonal ecological pattern. In the true hunting districts (i.e., the N, NE, SE, and NW regions) the catch of seals and whales is the most important source of subsistence. About 20% of the Greenlandic population lives in these regions.

Not only do catches vary from one year to the next, but long-term fluctuations can be demonstrated for a number of species. In some cases the possible reason for these fluctuations (climate, hunting pressure, etc.) have been indicated, but a more detailed discussion of these matters can be found in the literature (e.g., Anon. 1944, Bræstrup 1941, Vibe 1967, Hansen and Hermann 1953).

Employment Opportunities/Alternatives

Originally, hunting was the most common occupation and sole basis of survival in Greenland. The relative importance of various prey species varied with time and season and, to some degree, from locality to locality. Similarly, settlement patterns changed with variations in hunting conditions. Detailed accounts of seasonal variations in Greenland and their importance have previously been provided by several authors (e.g., Amdrup 1921, Anon. 1944, Rosendal 1958, 1961; Kapel 1975). During the 18th and 19th centuries this nomadic way of life gradually transformed into a more stable pattern with a couple of hundred small settlements and a few larger ones, although seasonal movements still took place. A wide variety of prey species were utilized, mainly seals (especially ringed and harp seals), but also whales, sea birds, caribou and others, depending on their seasonal or regional availability.

A change in this pattern appeared at the turn of the century and increased during the 1920's and 1930's. The seal hunt decreased drastically, especially in southwest Greenland, causing severe problems for the population in this part of the country. At the same time, cod was found to be commercially viable in the Davis Strait. These factors formed the background for an attempt to introduce a new pattern of occupation and commerce. A transition to fishing was coordinated and, carried out during the 1930s and 1940s, after which it evolved rapidly into the modern situation. No similar developments took place in northwest and east Greenland, although changes occurring in the southern part of the country were also felt in these regions.

Population policy in Greenland after World War II focused on development plans for the central and southern parts of West Greenland, and migration to this area was encouraged in different ways. The effect of this policy can be seen today as the total population in the hunting area proper is not much larger now than it was in 1900 or 1930. There are several reasons for this. The range of jobs is more diverse, and wages are generally higher in the fishery districts than in the hunting districts. The level of education is also normally lower in the small hunting communities. Young people from the hunting areas are often sent away to be educated in the central area, or even in Denmark. In this way, they miss their training in hunting at the most crucial stage of their life, and many of them have difficulty re-adapting themselves into the cooperative pattern of the hunting communities. This loss results in the migration of youth into the population centres. Although this may appear to offer some advantage to the hunting community, in that it helps avoid overpopulation, it may also result in the erosion of essential hunting culture values, and a transfer of small community problems into the larger communities where there is already severe unemployment, particularly for young people. Thus, the migration of people from the small hunting communities to the larger centres produces cultural and social problems in both places. The policy of population concentration realized in the previous decades has, therefore, in recent years provoked severe criticism, not least because of its adverse effects in the hunting areas. A breakdown of the population of Greenland by age, gender, and birthplace is seen in Figure 2.

Growing unemployment is today one of the most serious social problems. People, who come from hunting and fishing occupations, have very few chances to find jobs. There have been severe problems in determining the number of unemployed persons. This is partly because unemployment in some places is seasonal, partly because many people can have several types of small jobs at the same time (e.g., fishing, hunting, wage-earning in the fishing industry, domestic, industrial), and still do not earn enough to live.

Details of the Hunt: Historical Review, by Species

As described in the previous section, the hunting of large whales has always been part of a complex subsistence pattern which includes the utilization of other marine resources such as seals (five different species), walrus, polar bear, small cetaceans (five different species), sea birds (more than 20 species caught in significant numbers), and several fish species. In addition, hunting of some terrestrial mammals (caribou, muskox and Arctic fox) and birds has played a significant role in some regions at certain times.

P. Helms, O. Hertz, F.O. Kapel

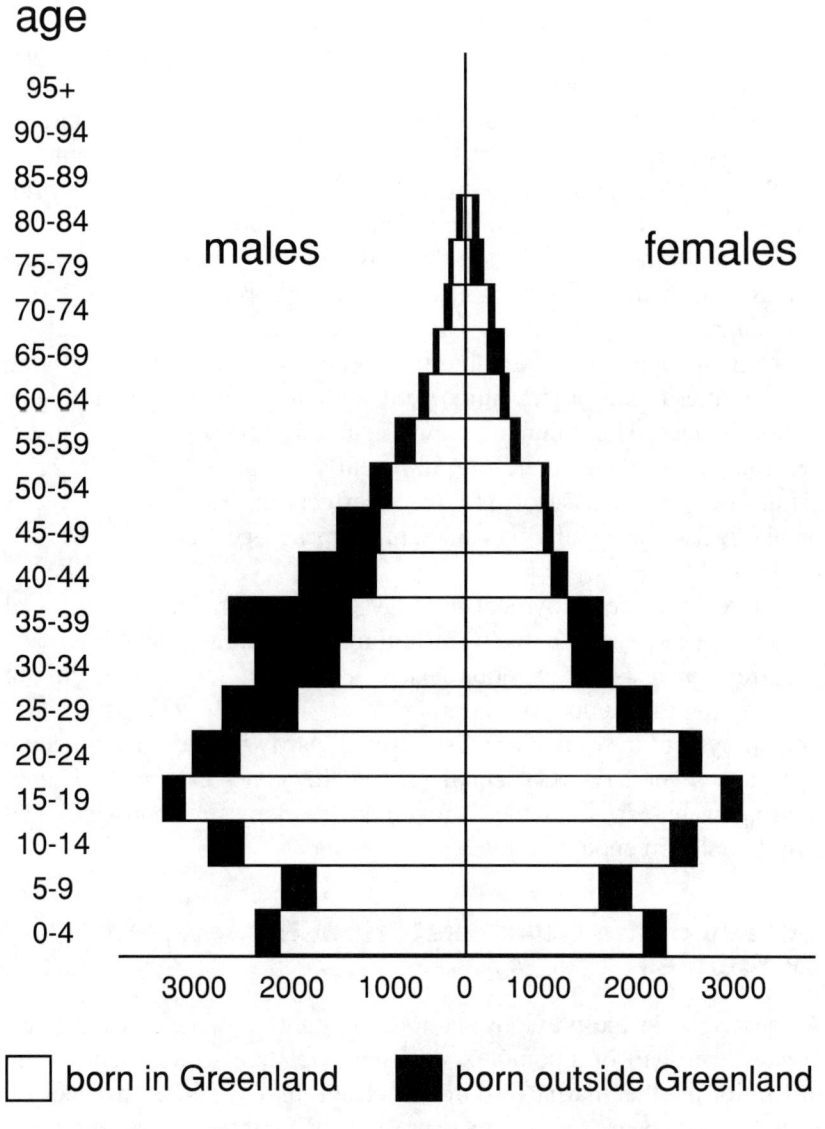

Figure 2. Demographic profile for total Greenland population, 1 January 1982

Regional variation in respect to the relative importance of the various species exploited in the pattern of subsistence was described in the previous section. As an introduction to the description of whale hunting given below, a short review of the other elements is, however, considered necessary.

Seal Hunting

The present take of seal in Greenland is estimated at 80,000 to 120,000 per year. For comparison, the total take in the middle of the 19th century was estimated at 89,000 seals (Rink 1877 [1974]). The total take, and the relative contribution of the various species is subject to significant annual and long-term fluctuations. The reasons for these fluctuations has been explained by long-term climatic variation (e.g., Vibe 1967) or by over-exploitation at the breeding patches outside Greenland (also Sergeant 1976). Recently, ringed seal accounts for the major part of the seal catch in Greenland (79-88%), with harp seal being the second most important species (7-13%). According to Rink the relative contribution of ringed and harp seals to the overall catch in the 19th century was 57% and 26%, respectively. This difference may be explained in part by the importance of seal hunting in Southwest Greenland at that time, but may also reflect long-term fluctuations in the abundance of the two species.

Walrus and Polar Bear Hunting

The present catch of walrus in Greenland is at the level of 200 to 300 animals per year, most of which are taken in North Greenland (Thule district). The species was previously an important part of the hunt in Central West Greenland, but today only a few dozen are taken in this and other regions. The polar bear is particularly important in North Greenland and East Greenland, whereas only single specimens are caught in other regions. The annual take is around 100 to 200 bears.

Hunting of Small Cetaceans

The total annual take of small whales is at the level of 2,000 to 3,000. Harbour porpoise in the southwestern part of Greenland, and narwhal and beluga in the northern and eastern parts of the country make up most of the catch. The take of narwhal and beluga is subject to large local variations from one year to the next, but is apparently at approximately the same level as one hundred years ago (Rink 1877 [1974]). The catch of harbour porpoise, of little importance in Rink's time, evolved in the early part of the 20th century, and has fluctuated between 500 and 1,500 animals per year in the most recent decades. The catch of marine mammals in various regions between 1978 and 1983 (provisional) is provided in Table 2.

Caribou Hunting

The caribou is presently found in West Greenland from the inner part of the Disko Bay (69°N) to the Frederikshab district (62°N), with small isolated groups on the Nuussuaq and Svartenhuk peninsulas, and on the Disko Island. The total population of caribou in West Greenland fluctuates considerably. The species was abundant from 1820-60 and, again, from 1900-20, but decreased to a very low level in the 1930's and 1940's. It recovered during the 1950's and reached a maximum in the late 1960's and early 1970's. A recent study estimates that caribou numbered about 100,000 at that time and suggests that the present 'stock' may be as low as 20,000-30,000 animals.

These fluctuations are, of course, reflected in the catches which also show yearly variations. In the middle of the 19th century catches of more than 25,000 were taken annually, whereas only a few thousand were taken between 1920 and the early 1950's. During the 1960's, the catches increased from 4,000 to 10,000 per year and peaked at more than 15,000 in the early 1970's. In recent years the annual catch has been at the level of 5,000 to 7,000.

Muskox Hunting

The muskox is presently found in northeastern Greenland from Peary Land to the region just north of Scoresbysund. This population has been subject to great fluctuations, primarily owing to changing climatic conditions. From a peak of 15,000 in the 1920's and 1930's, the population decreased to approximately 5,000 animals in the 1940's and 1950's. Since then the population has been recovering, and is at present estimated to be between 6,000 to 12,000 animals. The only hunting of muskox in Greenland is carried out by hunters from Scoresbysund, who take about 100 animals annually.

Fox Hunting

The Arctic fox is found all over Greenland. The size of the population is not known, but is subject to fluctuations as reflected in the catches. In the period 1795 — 1939 the number of fox skins traded in West

The Greenland Aboriginal Whale Hunt

Table 2.
Catch of Marine Mammals in Different Regions of Greenland, 1978 to 1983.

	1978	1979	1980	1981	1982	1983
North & Northwest						
Seals (Pinnipeds)	65,174	73,099	54,877	45,143	48,237	-
Small cetaceans	522	571	660	859	861	371[*]
Minke whales	30	38	20	15	17	25
Fins & Humpbacks	-	-	1	1	1	-
Central West						
Seals (Pinnipeds)	17,482	19,578	16,602	23,555	-	-
Small cetaceans	783	682	657	702	511	152[*]
Minke whales	50	58	64	59	71	66[*]
Fins & Humpbacks	1	-	6	5	5	1
South West						
Seals (Pinnipeds)	5,433	4,515	3,999	5,473	-	-
Small cetaceans	907	970	1,223	1,050	834	723[*]
Minke whales	58	83	99	96	14	60[*]
Fins & Humpbacks	30	21	19	13	15	20
South						
Seals (Pinnipeds)	4,162	3,184	3,216	3,277	-	-
Small cetaceans	19	38	47	16	5	1[*]
Minke whales	42	71	75	34	48	58
Fins & Humpbacks	-	-	2	-	-	-
South East & North East						
Seals (Pinnipeds)	16,579	16,872	15,844	20,786	-	-
Small cetaceans	6	21	58	144	104	41[*]
Minke whales	-	-	2	-	1	9[*]
Fins & Humpbacks	-	-	-	-	-	-

[*] Provisional estimates.

Greenland showed a general increase from around 1,000 to 4,000 annually. In Southwest Greenland peak periods occurred around 1830, 1875, 1920 and 1930, when more than 3,000 skins were traded annually. To the above mentioned figures should be added skins traded in the Thule district (fluctuating between 500 and 2,000 annually) and in East Greenland (50 to 300 per year) since the beginning of the 20th century. The present level of fox skins traded is 1,700 to 2,400 for all of Greenland.

Bird Hunting

The number of sea-birds caught in Greenland is not known exactly, but significant numbers are taken and used locally. A recent estimate places the total take at 500,000 to 1,000,000 annually, most of which are thickbilled murre and eider ducks.

Fishing

Although this section is meant to give details on hunting, it should be kept in mind that for the major part of Greenland's population, fishing is today more economically important than hunting, and these activities are often intimately connected with each other. A short review of the historic development of fishing is, therefore, relevant.

In ancient Greenland, fishing was carried out only as a minor supplement to hunting. Capelin, Arctic char, Greenland shark, and to a lesser extent Greenland halibut and some other fish species were caught seasonally for domestic use, which included dog feed. This local use is still a part of the subsistence pattern in the hunting districts.

Export of fish products started in the late 19th century, when small quantities of Arctic char and Greenland halibut were salted and sent to Denmark, but it was not until the beginning of the 20th century, that a commercial fishery really began. The importance of this fishery grew slowly until 1925, thereafter more rapidly, especially in the 1940's and 1950's. The last development has been a gradual transition from a mainly inshore fishery towards increasing participation in the offshore fishery. In recent years the most important species have been cod, deepwater prawn, Atlantic salmon, and Greenland halibut. All these fisheries are now under quota regulation. Further developments of the Greenlandic fishery involves the building up of a competitive fishing industry, the continuing assessment of resources, and international negotiations. In the years 1979-81 the Greenland catch amounted to 47,000-49,000 tons of cod, 20,000-40,000 tons of deepwater prawn, 1,200-1,400 tons of salmon, 5,300-5,800 tons of Greenland halibut, 3,600-5,600 tons of uvak, 1,700-4,000 tons of wolfish, and a small amount of other species.

Hunting of Large Whales

The bowhead whale was hunted from skin-covered boats (*umiat*) as an integral part of the hunting culture before European whaling began in the Arctic. After the Danish-Norwegian colonization of West Greenland in 1721, aboriginal bowhead whaling continued, often in cooperation with the colonists' shore-based whaling. As a result of commercial whaling in Davis Strait and Baffin Bay, the bowhead population was heavily reduced, and by the middle of the 19th century the Greenland bowhead hunt lost its importance. Few whales were caught in the remaining part of that century (Rink 1877 [1974], Winge 1902), and in the early part of the 20th century, only one or two specimens were taken (Kapel 1979, Mitchell and Reeves 1982). The Greenland authorities have agreed to give this species protected status, and only one dispensation from this regulation has been issued (in 1973). The bowhead whale is regularly seen in some areas of Greenland (Born and Heide-Jørgensen 1983), but it is no longer hunted. An exceptional and incidental net-entanglement of a young bowhead whale in North Greenland occurred, however, in November of 1980 (Kapel 1984).

The humpback whale was originally caught from umiat (pl.) in some regions of Greenland, especially near Frederikshåb and Gødthab, Southwest Greenland. The whales were approached when sleeping at the surface, harpooned and killed by lances (Fabricius 1780 [1929], Fabricius 1818, Rink 1877 [1974]). At a later stage, the boats and equipment used by the colonists for bowhead were occasionally placed at the disposal of Greenlanders for their catch of humpbacks. This activity was carried out continuously at Frederikshåb and Gødthab until 1923 (Winge 1902, Anon. 1944a, Møller 1971).

The traditional hunting method was discontinued, when modern catcherboat whaling was introduced by the Greenland Office in 1924. From this operation, the meat and intestinal fat was given to the local people for their consumption and the feeding of sled dogs (Anon. 1944b). The main target of the whaling operations was the fin whale, but humpbacks were also taken in small numbers until 1953 (Kapel 1979).

In 1948 the first Greenlandic fishing vessel was equipped with a harpoon cannon, and since 1958, when the above mentioned whaling vessel ceased operating, some fishing vessels engaged in minke whale hunting have occasionally taken a few humpback whales. The vessels catching humpback whales change from one season to the next, and none of them has caught humpbacks in every season. Most of the humpback whales were caught in Southwest Greenland, but some have been taken in Central West or Northwest Greenland. The total annual catch varied between 0 and 17 from 1958 to 1977 (Kapel 1979, 1983).

The fin whale was originally not hunted in Greenland, although dead animals were occasionally utilized (Rink 1877 [1974]). As mentioned above, this species was the main target of the whale-catcher operations working in Greenland in the periods 1924-39 and 1946-58. One purpose of these operations was to provide Greenlanders with meat, and for this reason the whales were landed at various settlements where local people received the meat and intestinal fat free in exchange for assisting during the flensing. The annual catch by the vessel averaged 23, but varied considerably from 3 to 51 (Kapel 1979).

Some of the fishing vessels that began hunting minke whale in 1948 have occasionally taken fin whales (0-13 per year until 1977). As in the case of humpback whaling by these vessels, the vessels actually hunting fin whales have changed from one season to the next, even more so for fin whales, because the catches were more evenly distributed along the coast of West Greenland (Kapel 1979).

Blue and sei whales were taken in small numbers by the whale-catcher operating in West Greenland between 1924 and 1958. Totals for the entire period were 44 and 8, respectively. A few whales caught by the fishing vessels since 1948 were initially reported as 'blue' or 'sei' (3 and 15, respectively), but there are reasons to believe that most of these were actually fin whales (Kapel 1984).

Exploitation of the minke whale in West Greenland began in 1948, when the first fishing vessel equipped with a harpoon cannon made this type of hunting possible, but remained at a low level during the following decade (avg. = 18 animals per season). During the 1960's, the total catch increased from approximately 50 to 200-300 animals per year, because an increasing number of fishing vessels were equipped with harpoon cannons and participated in the taking of minke whales on an opportunistic basis. In 1977 it was estimated that the total number of fishing vessels equipped with a harpoon cannon was around 100, but that only half of them were active in any given season (Kapel 1978).

In the early 1970's, an alternative method of catching minke, the collective hunting from a number of small boats with outboards, was introduced in some districts. Around 1975 it was estimated that approximately 18% (4-33%) of the total catch of minke whales in West Greenland was taken by this collective hunting method (Kapel 1978). The total catch averaged 233 minke whales annually in the period from 1973 to 82 (min.=180, max.=285).

Organization and Seasonal Variation in the Hunt of Large Whales

Since the cessation of the catcher-boat operations in 1958, minke, fin and humpback whales have only been hunted in Greenland by local people only for subsistence purposes.

The organization of the hunt of minke whale was reviewed in 1977 (Kapel 1978). It was found that around 125 fishing vessels had reported a catch of one or more minke whales since 1948. Some of the vessels had left the fishery during the period in question, and it was estimated that approximately 100 vessels were equipped with a harpoon cannon in 1976, with only about half of them (30 to 56) reporting catches in any one year between 1965-76. It should be stressed that, with few exceptions, all these vessels are primarily fishing vessels, which only occasionally engage in whale hunting. They may go to sea with the latter purpose in mind when whales have been observed in the neighbourhood, or they may change their mind at sea and shift from fishing to whaling as opportunity arises. But fishing is the *raison d'être* of the voyage.

The number of whales caught represents a minimum for the number of hunts, as no more than one whale is ever taken during a trip. The number of unsuccessful hunting attempts are not reported, so exact data on this aspect of the activity are lacking. Because of the opportunistic character of the fishery it may also, in practice, be difficult to classify a given trip as either a fishing or a whaling voyage. The size of the vessels engaged in whale hunting varies considerably (8-66 gross tonnage), but most of them are in the order of 15-30 tons.

As mentioned above, some of these vessels occasionally take a fin or a humpback whale. In the period between 1952-79 about 40 different vessels reported the catch of one or another of these species, but none have done so on a continuous basis. The maximum number of vessels taking fin or humpback whales in one season is 12, and the maximum number of whales taken by any one vessel in a given season is two for fin whales and five for humpback whales.

The seasonal distribution of catches varies between districts and to some degree from one season to the next. Generally, the catches of minke whale peaks in late spring (May) and autumn (October) in Southwest Greenland, in summer and autumn (June-October) in Central West Greenland, and in late summer (July-September) in Northwest Greenland. The catches of fin and humpback whales are distributed over the summer and autumn (June-October) in a similar way, with single specimens taken at other times. In recent years most humpback whales were taken in early summer (May-June), while fin whales were taken later in the year (September-October).

Review of the 1978-82 and 1983 Seasons

The hunting of marine mammals in Greenland from 1978-82 did not change significantly from the general pattern described above. Hunting statistics are summarized in Table 2. The number of seals taken varied between approximately 95,000 and 117,000 (all species), but showed no obvious trend during the period, and the regional distribution of catches remained on the whole the same as previously. The same is true for the catch of small cetaceans (n= 2,200 to 2,800), although the variations in catches during the period was not quite the same for all species in the various regions.

This implies that the relative importance of the catch of minke, fin and humpback whales remained unchanged during the period in question. The number of minke, fin, and humpback whales caught per district in each of the years between 1978 and 1982, and provisional figures for 1983, are shown in Table 3. Catch statistics for 1983 are not yet available for seals (and, for 1982, only for some districts). For small cetaceans up-to-date information is also lacking for some districts, so the figures in Table 2 should be considered preliminary for these years.

Seasonal Variation in the hunt of whales 1978-82/83
The seasonal distribution of whale catches appears in Table 4. As in previous years, most minke whales were taken between late spring and the autumn, with some variation between years and regions in respect to peak periods. The majority of fin whales were taken in the autumn, in 1978-79 and 1983 particularly in Southwest Greenland; in 1980-82 to a great extent in Central West Greenland.

Table 3.
Number of Minke, Fin, and Humpback Whales Caught in Greenland, 1978-82 and 1983 (provisional).

District	Minke						Fin						Humpback						Population (1982)	
	78	79	80	81	82	83[1]	78	79	80	81	82	83[1]	78	79	80	81	82	83[1]	Settlements/Inhabitants[5]	
Thule	–	–	–	–	–	–	–	–	–	–	–	–	–	–	–	–	–	–	6	720
Upernavik	8	8	3	4	3	2	–	–	–	–	–	–	–	–	–	–	1	–	11	1,992
Umanaq	22	30	17	11	14	23	–	–	–	1	1	–	–	–	1	–	–	–	8	2,221
Jakobshavn	10	1	10	8	5	–[2]	–	–	1	1	1	1	–	–	–	–	–	–	5	3,704
Christianshåb	11	11	10	8	9	1[2]	–	–	3	1	3	–	–	–	–	–	–	–	2	1,713
Godhavn	12	10	8	22	21	27	–	–	2	3	1	–	–	–	–	–	–	2	944	3,000
Egedesminde	15	35	36	21	31	27	1	–	–	–	–	–	–	–	–	–	–	–	3	1,152
Kangaatsiaq	2	1	–	–	5	11	–	–	–	–	–	–	–	–	–	–	–	–	5	
Holsteinsborg	11	15	10	5	13	–[2]	–	–	–	–	–	–	–	–	–	1	1	–	3	3,797
Sukkertoppen	18	40	52	43	52	39	3	4	–	–	–	2	5	3	3	5	2	3	4	3,448
Godthåb	18	21	26	33	33	21	1	–	3	–	2	2	13	10	5	2	7	6	3 (5)[6]	7,053
Frederikshåb	11	7	11	15	16	–[2]	3	3	3	1	1	2	5	1	5	4	2	5	4	2,379
Narssaq	4	–	3	–	2	4	–	–	–	–	–	–	–	–	–	–	–	–	3(10)[6]	1,698
Julianehåb	25	27	27	18	22	38	–	–	–	–	–	–	–	–	1	–	–	–	5(13)6	2,596
Nanortalik	3	44	45	16	24	16	–	–	–	–	–	–	–	–	–	–	–	–	7(13)6	2,586
Sum West Greenland	180	250	258	204	250	209	8	7	13	7	9	7	23	14	15	12	12	14	71(94)	39,003
				(250[3])														(15[4])		
Angmagssalik	–	–	–	–	1	–[2]	–	–	–	–	–	–	–	–	–	–	–	–	8	2,513
Scoresbysund	–	–	2	–	–	9	–	–	–	–	–	–	–	–	–	–	–	–	3	443
Sum East Greenland	–	–	2	–	1	9	–	–	–	–	–	–	–	–	–	–	–	–	11	2,956
																			[82 (105)	41,959]

[1] Provisional figures. [2] Reports not yet received. [3] Including estimates for not reported catches. [4] Including one specimen entangled in fishing net.
[5] Only residents born in Greenland. [6] Including settlements with less than 15 inhabitants (sheep farms etc.).

Table 4.
Seasonal Distribution of Large Whale Catches in Greenland, 1978 to 1983.

	Jan.	Feb.	Mar.	Apr.	May	June	July	Aug.	Sept.	Oct.	Nov.	Dec.	Season Unknown
Minke													
1978	-	-	-	13	17	18	22	18	25	16	6	7	2
1979	-	-	3	15	20	28	32	27	33	23	6	-	2
1980	-	-	-	13	23	43	31	15	33	35	11	1	55
1981	-	2	-	10	18	26	22	18	33	27	12	1	35
1982	-	-	2	9	17	37	14	16	50	37	9	-	12
1983	1	-	-	2	19	21	37	47	34	29	7	-	19
Fin													
1978	-	1	-	-	-	2	-	-	2	1	1	1	-
1979	-	-	-	-	-	-	-	2	2	2	-	1	-
1980	-	-	-	-	-	-	-	2	-	3	1	7	13
1981	-	-	-	-	-	-	1	-	2	2	-	-	2
1982	-	-	-	-	-	-	1	-	5	3	-	-	-
1983	-	-	-	-	-	-	-	-	4	3	-	-	-
Humpback													
1978	-	1	-	-	-	3	1	1	6	7	2	-	-
1979	2	1	-	-	1	3	2	2	2	1	-	-	-
1980	-	-	-	2	3	2	4	1	-	-	-	-	3
1981	-	-	-	2	4	5	-	1	-	-	-	-	-
1982	1	-	-	1	4	6	-	-	-	-	-	-	-
1983	-	-	-	-	4	9	-	-	-	1	-	-	-

Whale Hunting Methods and Efficiency

The size of the fishing vessels engaged in the hunting of minke, fin and humpback whales in Greenland average 15 to 30 gross register tons, some smaller than that, a few larger. The harpoon cannon is most often of Norwegian manufacture, calibre 50 mm, and the harpoon is attached to a heavy rope, in present days most often made of spun nylon. The harpoon-head is non-explosive. The equipment and the hunting technique is thus very much like that used in Norwegian small-type whaling.

In the collective hunting method, minke whales are secured by the use of hand-harpoons and floating bladders, and then killed by high-power rifles. The coordination of the participating small boats is often facilitated by the use of walkie-talkies, and sometimes a leader of the operation is appointed beforehand. The method has the advantage that it does not require the investment in expensive equipment associated with large vessels and harpoon cannons, which otherwise would be a doubtful and economically risky enterprise in the small settlements and the hunting districts.

The number of successful hunts is reflected by the number of whales taken. There is no new information on unsuccessful hunts, or the number of whales struck and lost. For this reason the question on hunting efficiency cannot be answered in any quantitative way at present. Quantitatively, the hunting methods used meet local needs, and thus are thought to be efficient.

Cultural Needs

Participation in Preparations for the Whale Hunt, the Hunt Itself, and Processing the Products of the Hunt

As mentioned above, we have no detailed information about the exact number of people participating in the whale hunt. Normally it is the same people who prepare for the hunt that participate in it. Where whaling is carried out from a fishing boat with harpoon guns, the crew is often composed of three to four men. When whaling is carried out from small boats crews average one to two persons, and local rules prescribe that not less than five boats may hunt a minke whale in order to ensure that the whale will not sink. Often 10 to 20 small boats take part in the hunt.

After a large whale is killed, it is towed to shore at high tide to be flensed at low tide. Sometimes a whale is flensed far from the town or settlement. At other times it is towed home, where others can assist hauling the whale ashore by means of a pulley. In a smaller village nearly all adults will help in this work. All those helping can take a share of the meat and the whale skin (*muktuk*), and noone will be denied cutting pieces of *muktuk* to be eaten raw on the spot, to the great pleasure of the children.

Distribution and Sharing of Whale Meat and Products, Food Preferences, and Degree of Utilization

Rules for dividing whales differ from place to place and are dependent on the species of whale concerned. A minke whale hunted by open boats will, in the Umanaq-district, be equally divided between all participating boats. If it is hunted by a fishing boat with harpoon gun, the boat will get one part, the owner/skipper one, and the crew the rest. When one boat with harpoon gun and several open boats have cooperated, it sometimes raises difficult problems in finding an equitable way of dividing the meat that everyone present will find acceptable. In *Scoresbysund*, on the east coast, the hunt of minke whales is totally municipal and the meat is divided and provided free to all people in the local community. In dividing smaller whales, such as narwhal and beluga, different rules are used according to what type of boat is used, or on the hunter's decision to use old sharing rules or newer ones. The old rules can be very complicated, as the hunting-shares are dependent on who first harpooned/shot the whale, who did it next, and so on.

In the smaller settlements, and to a certain degree in towns, the meat a hunter provides will be further distributed to his family, friends, partners, or people he is related to through a special system of name sharing. Gifts of this sort are very highly prized, especially in towns where meat from seals and whales, the only 'real' sort of food, is scarce.

Sometimes whale meat is sold to The Royal Greenland Trade Department or to a local private fishing factory, which distributes it to shops along the coast.

Ceremonies, Feasts, and Folklore

In Greenlandic folklore several tales about whales and whaling are found. Stories of actual whaling events are told over and over again to hunters and their children, who in turn receive important information, knowledge and values. Ceremonies and feasts, such as in Alaska, are not found in Greenland today. But a ceremony takes place when a boy has caught his first narwhal or beluga whale. All friends and family are invited to a feast or given parts of the whale, while he himself may not eat. Formerly, certain rules were followed in the whale hunt, e.g., a whaler's clothes had to be clean, but today no special ceremonies take place before or after a hunt. A common public dancing feast can take place, with shanties and jigs learned from the old European and American whalers, who whaled commercially in the Davis Strait.

Social Integrative Functions of the Whale Hunt and Risk to Community Identity from Imposed Restrictions on Aboriginal Whaling

The social integrative functions of the whale hunt reflect community values and needs, and are necessary for cooperation and coordination at a local level. Cooperation in connection with the hunt, the taking of the whale, and its flensing ashore, followed by a distribution of meat and *muktuk* instills in the participants a feeling of mutual dependence as well as a feeling that people who need help will be helped. Participation in the hunt is also of great importance to a hunter's social prestige and sense of self-worth. Great prestige and feelings of self-sufficiency are attributed to hunting activities and especially to whaling throughout Greenland.

Whaling requires a great knowledge of whale ecology, weather, ice conditions, etc., and people who have acquired this knowledge have always been given special respect and prestige in Greenland's egalitarian communities. In areas where whale hunts take place there will often be one or more experienced people who are regarded as the local whaling specialists. These people act as leaders in whaling activities and are asked for decisions if uncertainty arises in connection with the sharing of the meat. They are usually elderly people, though not always.

The social importance of being a skilled hunter and whaler can also sometimes be seen in connection with municipal elections (such men often hold political positions). Whaling is, thus, intertwined with social life. If whaling ceased altogether, the need for whale meat would grow, especially in the smaller villages. However, the hunters' economic situation would worsen, and the socially important 'meat helping system' would be compromised.

Other Uses of Whales

Whales are hunted in Greenland only for domestic consumption, and the products from the whales are used intensively. In older days, these products found a greater diversity of uses than today, but the degree of utilization remains nearly the same. Except for some of the entrails, the skeleton and the baleen, every part of the whale is still used. The traditional and the present day utilization of whale products in Greenland are shown in Table 5.

Nutritional Needs

Role of Whale Products as Food in the Community and the Importance of Whale Products in the Traditional Diet.

Consumption of whale meat by species

Within the last five years (1977-81), whales have supplied an average 887 tonnes of meat per year in Greenland. Meat from the minke whale constitutes just under half this amount at 415 tonnes, followed by the beluga and humpback whale each contributing about 125 tonnes. The remaining species contribute less than 100 tonnes per year per species (Table 6). The 887 tonnes of whale meat correspond to 21.7 kg per year per capita born in Greenland (population 40,900), or .059 kg. per capita per day. A breakdown by species appears in Table 7.

No exact documentation is available on the contribution of whale meat from all species, as small whales are consumed by local groups, whereas meat and *muktuk* from large whales are distributed to a varying extent. From the Greenland catch records it is possible to determine which municipalities catch whales. The percentage of the total population having had the opportunity to consume whale meat has been calculated based on population figures in municipalities with regular whale catches (Table 7).

The number of persons born in Greenland has doubled between 1945 and 1982 to 42,000. This has necessitated increasing the importation of foods from Denmark. However, food balance sheets from Greenland have not been published, and the published figures of the monetary value of imports do not permit a breakdown on foods. A list of imported foods consumed in Angmagssalik, East Greenland (population 2,500) during 1978 forms the basis of the average breakdown of foods per capita diet per day in Table 8. The average consumption of Greenlandic large mammals appears from annual catch records, whereas the consumption

Table 5.
Examples of the Most Important Uses of Whale Products in Greenland.

Item	Use Today	Traditional Use
whale skin (*mataq*)	eaten raw or boiled, for sale	eaten raw or boiled, lashing thongs, whip crackers
blubber	eaten raw or boiled, for sale, sled dog food, smearing of lines and boots, fuel in emergency situations on hunting trips	eaten raw or boiled, for sale, sled dog food, smearing of lines and boots, kayaks, fuel in stone lamps
meat	eaten raw, boiled or dried, sled dog food, for sale	eaten raw, boiled or dried, sled dog food
blood	eaten raw, boiled or fermented	eaten raw, boiled or fermented
heart	eaten boiled	eaten boiled
sinew (small whale)	sewing thread	sewing thread, lashings
baleen	no use	fishing lines, snares, seal nets, lashings, bows, domestic utensils, figure carvings, support for bedding, trade/barter
teeth	harpoon heads, buttons, domestic industries, towing implements	harpoon heads, kayak implements, buttons, figure carvings, towing implements
tusk (narwhal)	domestic industries, for sale	foreshaft on harpoon, 'sucking raw', for sale
shoulder blade	no use	scraping board, shovel, house building
dorsal	no use	stool/chair
ribbon	snow trasher	snow trasher, snow knife, house building, sled runners

Table 6.
Catch of Large Mammals, 1977-81.
(Total consumption of whale meat by species)
(Per capita consumption of whalemeat by species)

	1977	1978	1979	1980	1981	Meat weight per head	Greenland total kg whale meat/year	Average meat weight per capita[1]	
						kg	kg	kg/year	g/day
HVALER I ALT (Total whales)	2,590	2,448	2,553	1,721	2,627				
heraf (incl.):									
Sildepisker, vagehval (Minke whale)	300	180	250	200	200	2,000	415,600	10.2	28
Narval (Narwhal)	274	615	395	250	600	225	84,105	2.1	6
Hvidhval (White whale)	805	719	741	400	900	200	125,240	3.1	8
Grind (Pilot whale)	136	101	50	50	-	400	26,880	0.7	2
Marsvin (Harbour porpoise)	1,007	798	1,075	800	900	20	18,320	0.4	1
Knølhval, pukkelhval (Humpback whale)	16	24	14	11	12	8,000	123,200	3.0	8
Finhval (Fin whale)	13	7	7	6	6	10,000	78,000	1.9	5
Andre (Others)	39	4	21	4	9	1,000[2]	15,400	0.4	1
Whale meat total							**886,745**	**21.7**	**59**

[1] Number of persons born in Greenland in 1977-1981 was about 40,900.
[2] Estimated value.

Table 7.
Percentage of Population Consuming Whale Meat by Species.

Species	% of Population
Minke whale	90
Narwhal	50
White whale	70
Pilot whale	50
Harbour porpoise	90
Humpback whale	40
Fin whale	20

of birds and fish are estimated. The total per capita diet per day in Table 8 supplies 10,000 kJ (2,400 kcal.), with only 1/4 of this energy being contributed by traditional Greenlandic food; 3/4 comes from imported processed food-types. Table 8 also shows that whale meat contributed .059 kg to the daily diet. This volume corresponds to 23% of the weight of Greenlandic meat products (whales, seals, reindeer, and birds) and to 17% of Greenlandic and Danish meat products (87 g). The breakdown of the .059 kg of whalemeat on species appears in Table 9, where the percentage contribution by species to total per capita diet has been computed by weight as well as energy value.

It appears from Table 10 that whale meat supplies 14% of the total volume of animal protein in the diet, whereas the contribution of whale meat to the total energy value is just under 5%. Therefore, whale meat is a major source of animal protein. Imported food types supply just under 20% of animal protein in the diet, but 74 % of the energy value. It is therefore evident that whale meat is an important source of maintaining the nutritional balance.

Alternative Food Resources to Whale Meat and Availability and Acceptability of Other Food Sources

The nutritional value of 60g of whale meat and 60g of seal meat is approximately the same. Theoretically, it would therefore be conceivable, in terms of nutritional value, to use seal meat instead of whale meat in all regions of Greenland, where it is the staple food *par excellence*.

Table 8.
Food Balance Sheets of Greenland.
(Detailed breakdown of total *per capita* diet per day)

FOOD ITEM Ave. consumption/day	Netto g	Energy kJ	Fat g	Fatty Acids satur- ated/g	Fatty Acids poly- unsat./g	Carbo- hydrate g	Starch g	Sugars g	Protein g
Whale Meat	59	475	5.9	1.2	0.8	0	0	0	14.8
Seal Meat	165	1,328	16.5	2.6	3.6	0	0	0	41.3
Caribou Meat	21	188	3.0	1.3	0.1	0	0	0	4.3
Birds	37*	295	4.4	1.4	1.0	0	0	0	7.7
Fish	110*	354	2.1	0.4	0.9	0	0	0	16.3
Total Greenlandic Foods	**392**	**2,640**	**31.9**	**6.9**	**6.4**	**0**	**0**	**0**	**84.4**
Dairy Products	94	346	4.9	3.0	0.1	4.4	0.0	0.0	4.9
Fats	22	689	18.0	7.2	3.1	0.1	0.0	0.1	0.1
Meat (87 g), EG (7 g)	94	856	15.3	5.9	2.1	0.1	0.0	0.1	16.0
Cereals & Cakes	188	2,766	13.6	4.4	3.5	114.7	100.1	14.6	17.7
Sugar	124	2,104	0.0	0.0	0.0	123.8	0.0	123.8	0.0
Potatoes	58	194	0.1	0.0	0.1	10.3	9.7	0.6	1.0
Fruits, vegetables, and preserves	119	407	0.2	0.0	0.0	22.5	0.3	22.2	1.1
Total Imported Foods	**699**	**7,362**	**52.1**	**20.5**	**8.9**	**275.9**	**110.1**	**165.8**	**40.8**
Grand Total	**1,091**	**10,002**	**84.0**	**27.4**	**15.3**	**275.9**	**110.1**	**165.8**	**125.2**
Energy			32			47		21	

* estimated values

Table 9.
Percentage Contribution of Whalemeat by Species to Total Per Capita Diet.

	Weight g/day	%	Energy kJ/day	%
Minke whale	28	2.6	225	2.3
Narwhal	6	0.5	48	0.5
White whale	8	0.7	65	0.8
Pilot whale	2	0.2	16	0.2
Harbour porpoise	1	0.1	8	0.1
Humpback whale	8	0.7	65	0.8
Fin whale	5	0.5	40	0.5
Others	1	0.1	8	0.1
Whale meat total	59	5.4	475	5.3
Total per capita diet	**1,091**	**100.0**	**10,002**	**100.0**

The question of whether it would actually be possible to replace part of the whale catch with seals is not considered here. It would theoretically be possible to replace whale meat by imported meat products with corresponding nutritional values, but the culinary or ritual values would not be the same. The Greenland population prefers meat from marine mammals to meat from livestock, especially pork and beef.

Any shift from the original Greenlandic diet to non-native foods would increase the risk of introducing diet and nutrition related diseases associated with western civilization into Greenland society. Such diseases are the result of refined foods from industrial food production processes. These foods are high in simple carbohydrates and fat content with a low ratio of polyunsaturated to saturated fatty acids. The recommended total diet should not contain more than 12g of simple carbohydrates per 10,000 kJ. The optimum ratio of polyunsaturated to saturated fatty acids is 1.0 or above.

The traditional Greenlandic and imported food types shown in Table 8 are compared per 1,000 kJ in Table 11. The latter table shows

Table 10.
Role of Whale Products as Animal Protein and as a Source of Calories

	Weight			Animal Protein			Energy		
	g/day	Grl. prod. (%)	Total (%)	g/day	Grl. prod (%)	Total (%)	g/day	Grl. prod (%)	Total (%)
Whale meat	59	15.1	5.4	14.8	17.5	14.1	475	18.0	4.7
Seal meat	165	42.1	15.1	41.3	49.0	39.2	1,328	50.3	13.3
Caribou	21	5.4	1.9	4.3	5.1	4.1	188	7.1	1.9
Birds	37	9.4	3.4	7.7	9.1	7.3	295	11.2	3.0
Fish	110	28.0	10.1	16.3	19.3	15.5	354	13.4	3.5
Greenlandic products	392	100.0	35.9	84.4	100.0	80.2	2,640	100.0	26.4
Imported food items	699		64.1	20.9		19.8	7,362		73.6
Total	1,091		100.0	105.3		100.0	10,002		100.0

that Greenland imports refined processed food types content of 22.5g simple carbohydrates/1,000 kJ. The immediate consequence of a diet composed largely of these foods is an extremely high incidence of dental caries which was non existent in the traditional Greenlandic diet. The ratio of polyunsaturated (1.2g) to saturated (2.8g) fatty acids in imported foods is considerably lower (0.4) than the corresponding ratio of 0.9 (2.4:2.6) in the Greenlandic diet. The high figure applying to fats in Greenlandic products seems to account for the much lower incidence of arteriosclerosis and coronary heart diseases in Greenland than in industrialized countries. The official propaganda for better food habits in Greenland heavily stresses the maintenance and increase of the use of Greenlandic products to prevent adverse effects of a shift to a non-native food diet.

Table 11.
Comparison of Traditional Greenlandic Foods with Imported Food Types.

	Per 1,000 kJ		Fatty Acids		Carbohydrates		
	Energy	Fat	Saturated	Polyunsaturated	Starch	Sugars	Protein
	Kj	g	g	g	g	g	g
Greenlandic food	1,000	12.1	2.6	2.4	0	0	32.0
Imported food	1,000	7.0	2.8	1.2	15.0	22.5	5.5

References

Ackman, R.G. and S.N. Hooper. 1974. Long-chain monoethylenic and other fatty acids in heart, liver, and blubber lipids of two harbour seals (*Phoca vitulina*) and one grey seal (*Halichoerus grypus*). *J. Fish. Res. Board Can.* 31:333-341.

Ackman, R.G., C.A. Eaton, and P.M. Jangaard. 1965. Lipids of the fin whale (*Balaenoptera physalus*) from North Atlantic waters. I: Fatty acid compositon of whole blubber and blubber sections. II: Fatty acid composition of the liver lipids and gas-liquid chromatographic evidence for the occurrence of 5, 8, 11, 14 nonadecatetraenoic acid. *Canadian Journal of Biochemistry* 43:1513, 1521.

Ackman, R.G., S. Epstein, and C.A. Eaton. 1971. Differences in the fatty acid compositions of blubber fats from northwestern Atlantic fin-whales (*Balaenoptera physalus*) and harp seals (*Pagophilus groenlandica*). *Comp. Biochem. Physiol.* 40B: 83-697.

Amdrup, G.C. et al. (eds). 1921. Grønland i Tohundredeåret for Hans Egedes Landing I-II. *Medd. om Grønl.*

Anonymous. 1944. Statistiske oplysninger om Grønland, III. Beretn. vedr. Grønl.

Anonymous. 1944a. Grønlændernes fangst af Søpattedyr: Knølhvalen. Sammendrag af Statistiske Medd. vedrørende Grønland III. *Beret. vedr. Grønlands Styrelse* No. 1: 627.

Anonymous. 1944b. Sødyrfangst under europæisk Ledelse. *Beret. vedr. Grønlands Styrelse* No. 1: 628.

Anonymous. 1977. En undersøgelse af indkomsterne i Upernavik bygder 1975 og en sammenligning med bygderne i Umanaq kommune. Grønlandsradets dok.nr. 32/77.

Anonymous. 1977-81. Sammendrag af Grønlands fangstlister m.v. Minist. for Grønl.

Anonymous. 1983. Leading article: Eskimo diets and diseases. *Lancet* I: 1139-1141.

Bang, H.O. and J. Dyerberg. 1975. Blodets fedtindhold og kostens sammensætning hos en vestgrønlandsk befolkningsgruppe. *Ugeskr. Læger* 137: 1641-1646.

Bang, H.O. and J. Dyerberg. 1981. The Lipid Metabolism in Greenlanders. *Meddelelser om Grønland, Man & Society* 2: 3-18.

Bang, H.O., J. Dyerberg, and H.M. Sinclair. 1980. The composition of the Eskimo food in northwestern Greenland. *Arn. Journ. Clin. Nutr.* 33: 2657-2661.

Bang, H.O., J. Dyerberg, and N. Hjørne. 1976. The composition of food consumed by Greenland Eskimos. *Acta Med. Scand.* 200:69-73.

Born, E.W. 1983. *Havpattedyr og havfugle i Scoresbysund.* Danbiu, APS.

Born, E.W. and M.P. Heide-Jørgensen. 1983. Observations of the bowhead whale (*Balaena mysticetus*) in Central West Greenland in March-May, 1982. *Rep. Int. Whal. Commn.* 33:545-48.

Bræstrup, F.W. 1941. A study on the Arctic fox in Greenland. *Medd. om Grønland* 131.

Denmark 1976-82 Statistisk for Årborg, Danmarks Statistik, 1976, 77, 78, 79, 80, 81, 82.

Dyerberg, J. and H.O. Bang. 1978. Dietary fat and thrombosis. *Lancet* 1978 I:152.

Dyerberg, J. and H.O. Bang. 1982. Factors influencing morbidity of acute myocardial infarction in Greenlanders. *Circumpolar Health*, 5th International Symposium, Nordic Council for Arctic Medical Studies, Rep. Ser. 33: 300-303.

Dyerberg, J., H.O. Bang, E. Stoffersen, S. Moncada, and J.R. Vane. 1978. Eicosapentaenoic acid and prevention of thrombosis and atherosclerosis? *Lancet* II. 117- 119.

Fabricius, O. 1780 (1929). Fauna Groenlandica. *Grønl. Selsk. Skr.* VI, 168 pp.

Fabricius, O. 1818. Om Stubhvalen. *Da. Vidn. Selsk. Skr.* 6(1):63-83.

Goddard, V.R. and L. Goodal. 1959. *Fatty acids in animal and plant products.* Washington, DC.

Gullestrup, H., Sørensen, and Scwerdfeger. 1976. Udviklingen i Grønlands bygder. *Nyt fra samfundsvid.* 40.

Hansen, P.M. and F. Hermann. 1953. Fisken og Havet ved Grønland. *Skr. Danm. Fisk. Hav.*, 16, 128 pp.

Helms, P. 1980. *Kostvurderingstabeller.* Akademisk Forlag, København.

Helms, P. 1981. Kostundersøgelse i Angmagssalik. *Forskning/tusaut i Grønland* 1-2:10-14.
Helms, P. 1981. Angmassalingme nereissanik misigssuineq. *Forskning/tusaut i Grønland* 1-2:14-18.
Helms, P. 1982. Changes in disease and food patterns in Angmagssalik, 1949-1979. Circumpolar Health, 5th International Symposium, Nordic Council for Arctic Medical Studies, Rep. Ser. No. 33:243-251.
Hertz, O. 1977a. Ikerasarssuk — en boplads i Vestgrønland. *Nationalmuseet.*
Hertz, O. 1977b. En økologisk undersøgelse af minedriftens virkninger for fangerne i Uvkusigssat. Minist. for Grønl.
Hertz, O. 1985. Uummannaq — *ecology and survival in the Arctic* (in print).
Holtved, E. 1962. Otto Fabricius' ethnographical Works. *Medd. om Grønl.* 140, nr. 2.
Holtved, E. 1967. Contributions to Polar Eskimo Ethnography. *Medd. om Grønl.*
Kapel, F.O. 1975. Recent research on seals and sealhunting in Greenland. Rapp. P. (v). *Réun. Cons. Int. Explor. Mer.* 169: 462-478.
Kapel, F.O. 1978. Catch on minke whales by fishing vessels in West Greenland. *Rep. Int. Whal. Comm.* 28: 217-26.
Kapel, F.O. 1979. Exploitation of large whales in West Greenland in the 20th century. *Rep. Int. Whal. Comm.* 29: 197-214.
Kapel, F.O. 1983. Denmark (Greenland). Progress report on cetacean research, June 1981 to May 1982. *Rep. Int. Whal. Comm.* 33: 203-08.
Kapel, F.O. 1984a. A note on the entanglement of a bowhead whale (*Balaena mysticetus*) in Northwest Greenland, November 1980. IWC paper SC/36/PS 8.
Kapel, F.O. 1984b. On the occurrence of sei whale (*Balaenoptera borealis*) in West Greenland waters. IWC paper SC/36/Ba 2.
Kapel, F.O. and R. Petersen. 1982. Subsistence hunting: The Greenland case. *Rep. Int. Whal. Comm.* Special Issue, 4: 51-73.
Melgaard, J. 1983. Qajaa, en køkkenmøddi i dybfrost. *Nationalmuseets Arbejdsmark.*
Ministeriet for Grønland. 1976-81. Sammendrag af fangstlister m.v. 1976, 1977, 1978, 1979, 1980, 1981.
Ministeriet for Grønland. 1982. Årsberetning: Grønland 1982, 13.
Mitchell, E.D. and R.R. Reeves. 1982. Factors affecting abundance of bowhead whales (*Balaena mysticetus*) in the eastern Arctic of North America, 1915-1980. *Biol. Conserv.* 22: 59-78.
Møller, J. 1971. Knølhvalfangst. In: K. Hansen (ed.), *Grønlandske fanqere fortæller,* 53-62. Nord. Landes (1922-23) Bogforl. 1.
Petersen, R. n.d. Gamle og nye autoriteter i Grønland (særtryk).
Petersen, R. 1970. Tidsskrift for Grønl. *Retsvæsen.* 6, nr. 1.
Rasmussen, K. 1979. *Myter og Sagn I — III.* Kobenhavn.

Rink, H.J. 1877 *Danish Greenland, its people and products.* C. Hurst & Co., London; A. Busch, Copenhagen. (1974)
Rosendahl, P. 1958. Kårene i fangerdistrikterne. I-II. *Grønland* 58-68, 99-110.
Rosendahl, P. 1961. Grønlandsk jagt- og fangststatistik. Geogr. *Tidsskr.* 60:16-38.
Sergeant, D.E. 1976. History and present status of the population of harp and hooded seals. *Biol. Conserv.* 10: 95-118.
Vibe, C. 1950. The Marine Mammals. *Medd. om Grønl.* 150, nr. 6.
Vive, C. 1967. Arctic animals in relation to climatic fluctuations. *Medd. om Grønland* 170(5), 227 pp.
Winge, H. 1902. Grønlands Pattedyr. *Medd. om Grønland* 21(2): 317-521.

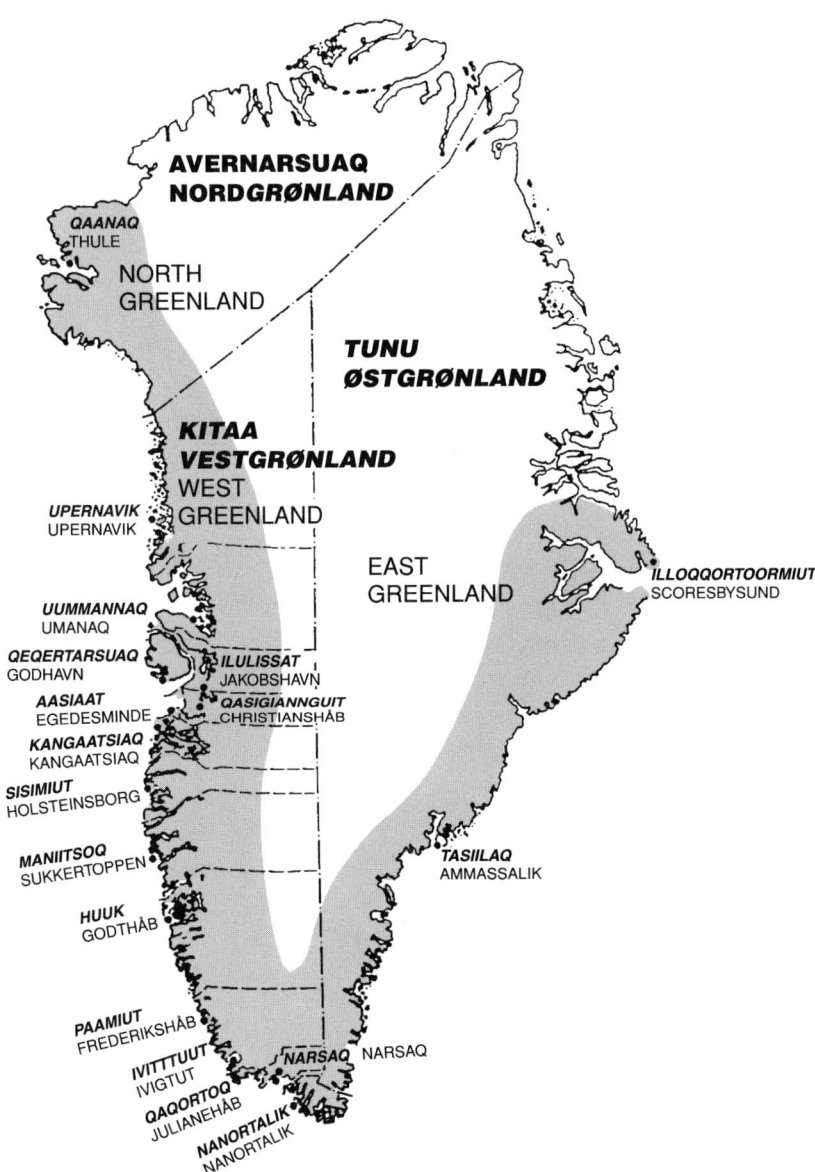

Figure 1. Subsistence Whaling in Greenland.

Communal Aspects of Preparing for Whaling, the Hunt Itself, and the Ensuing Products

R. Petersen
Institute of Eskimology
University of Copenhagen
1987

Introduction

It is difficult to assess in each community the exact number of people assisting in the preparatory phase of whaling. Whaling is simply a part of the whole of a profound dependence on living resources. The toothed whale hunt serves as a supplement to sealing, whereas baleen whales are normally taken from fishing vessels fitted with harpoon guns. The later type of whaling often assumes a random character, as it takes place when a fishing vessel happens to encounter a whale. Nevertheless, it remains an important part of Greenlandic livelihood. A fishing vessel of this kind may have a crew of 3 to 5 men who will be engaged in the hunt.

Distribution, Sharing, and the Role of Money

Formerly, the distribution of the products from whaling was a function of the communal nature of the resource base, but it also served as a kind of mutual insurance. This was clearly so in the egalitarian hunting communities of old. Everyone was expected to take part in the general distribution of the products from the whale hunt.

Today, people live in more economically diversified communities, and problems arise when families are unable to participate in the general distribution of whale meat or reciprocate within the framework of a pure subsistence economy. Often times, hunting products will be passed on unidirectionally from low-income to high-income groups. The hunters definitely make up a low-income group. A distribution system of so lopsided a character obviously cannot continue for long. Some type of compensatory element must be introduced into the system, and this is where money plays a role. Without financial compensation the channels of distribution would breakdown, and one of the most immediate effects would be higher rates of waste. Moreover, more people would feel the need to go hunting for themselves, which would put further pressure on the resource base (and personal time tables).

As is normally the case, meat from whale hunts is sold locally where money is incorporated into the distribution system. This means that, although some people are unable to take part in the hunt, they participate in the communal sharing of hunting products, thus obtaining their share. This does not necessarily signify any commercialization of the hunt *per se*. Rather, it is a means to keep the distribution channels open in a way that is beneficial to all parties. That way, at least one important part of the aboriginal system is maintained, namely the sharing of hunting products as some kind of mutual insurance.

It would, of course, be worrisome if money entered the system in such a way as to motivate people to invest in whaling and intensify it for personal gain or commercial profit. But there is actually no sign of whaling developing into anything else than what it has been for a long time, a supplementary form of Greenlandic natural resource extraction, tied especially to the fisheries.

Social Status, Prestige, and Community

In the whaling settlements of old, whalers were given a special status since whaling was a significant part of the economic basis of the community. Even though most whaling today is an ancillary activity related to the fisheries, it confers a certain status to and aura about a whaler. Whale products, especially meat and *muktuk,* are valuable and highly prized elements of the Greenlandic diet and are often served at festive occasions. Whale meat and *muktuk* thus take on significance for cementing social relations. At present, there are practically no financial incentives of any importance on the national level connected to whaling in Greenland. It is the cultural values that sustain and are the focus of the whale hunt.

Formerly, in certain areas of Greenland, settlements were built exclusively with whaling in mind, although sealing opportunities were the determining factor in the choice of the exact location. Whaling traditionally did much to bridge the gap between 'rich' and 'poor.' Not only did a landed whale provide food to the entire settlement, it also meant that less fortunate neighbouring communities could share in the bounty. Such was the recognized social importance of the communal aspects of whaling. Anyone without food, or a provider, had the right to help him- or herself to the meat and blubber brought in to the settlement, even if they, for one reason or another, were prevented from personally taking part in the flensing.

Even when smaller toothed whales were taken, the sharing rules provided a wider allowance for obtaining portions of the landed narwhals or beluga. These were the so called shares, to which all local hunters present were entitled, even if they themselves did not take part in the actual hunt. In those parts of Greenland, where smaller whales were commonly taken, not every hunter had to catch a whale to get some whale meat. These sharing rules were more or less the same throughout Greenland, and are applied to this day in areas where smaller whales are common.

There is social prestige attached to the landing of a whale. In the old days, transition from childhood to womanhood was determined largely by sexual maturity. In contrast, manhood was determined by cultural criteria and social recognition, such as the landing of a boy's first seal, or of course, whale. Even today, a boy's maturity is tied to the hunting activities of the family and is dependent on his ability as a bread-winner, or 'meat-winner,' to be exact. Even if confirmation in the church has taken on the role as "bridge between childhood and adulthood", landing of the first prey is still a significant rite of passage. Landing one's first mammal is very different from obtaining one's first paid job, even though this too has to do with the role as provider. There are no special ceremonies or festivities attached to the first money earned, but there always is to the first prey landed.

Not all meat landed is sold on the market. It is still common to give away part of the meat as free gifts to the elderly or to people to whom the hunter feels special bonds of friendship or gratitude. Especially in smaller settlements, people often receive 'meat-gifts' of different kinds if they have a job that prevents them from taking part in a hunt. The selling of meat for money takes place when the hunter deals with people with whom he has no normal or consistent sharing relationships. This corresponds to the bartering system of old, which took place between people who were not involved in a sharing relationship. The difference between that and the present-day situation is that today people who do not share their daily activities or values, or who live in different "worlds" so to speak, often times physically live as neighbours in the same town.

The system used for the distribution of resources has always differed from one part of Greenland to another. This, of course, was the *raison d'être* of bartering, which took place especially when people came together. People came in rowboats from all over for the bartering 'fairs' which were located in areas allowing for short-term, but intensive occupation. Bartering of this kind is possible now for only a very small segment of the population as people no longer have the same kind of small, inexpensive boats. A journey of the kind that was usual in earlier times would indeed be a very expensive proposition today if conducted on a modern standard vessel for long distance travel.

As it is, more out of necessity than choice, the great bartering 'fairs' of the past have been supplanted by the new routine whereby hunters bring their catch to the neighbouring town to sell for money instead of exchanging it for other goods. The hunters' meat products are then distributed widely through channels established by the Co-ops or by the Home Rule government.

Conclusion

Many communities in Greenland encounter difficulties providing a local year-round supply of 'strong meat' (i.e., meat from marine mammals). This inadequacy is definitely offset with aboriginal meat products channeled to communities from the hunting district proper. If this distribution was not allowed to take place, the alternative would be the consumption of foreign products such as chicken or New Zealand lamb. But this kind of meat is inferior to whale meat and simply does not contain the amount of calories and cholesterol-free fat needed to go out hunting and fishing in weather of -40° C.

The cultural needs of whale meat and *muktuk*, not to mention the meat of seals, caribou, and other animals, are as rich as they are diverse. It is little wonder that the sharing and consumption of these meats is the traditional high point of any festive occasion in Greenland.

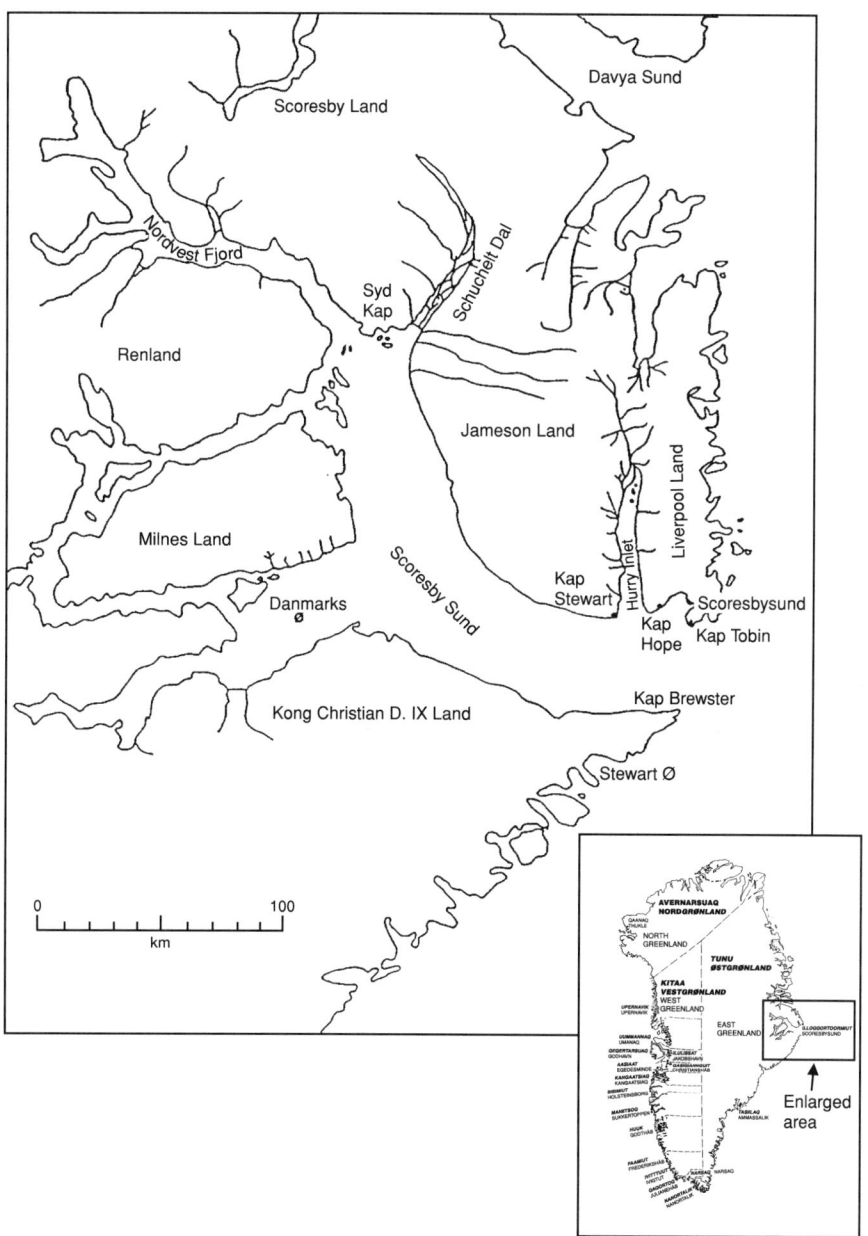

Figure 1. Map of Scoresbysund District.

Scoresbysund: A Hunting Community in East Greenland

Fin Brienholt Larsen
with Ivalo Egede and Carolyn J. Jenkins
Inuit Institute, Nuuk
1987

Introduction

Subsistence whaling forms an integral part of hunting activities in Scoresbysund. This paper presents a general description of the community, its people, and their use of the renewable resources in the area. The importance of whaling to the local population is evaluated within this general background.

Scoresbysund is located at the mouth of the longest fjord in the world, the Scoresbysund Fjord, which lies about 1,000 km north of Ammassalik. With the exception of remote meteorological and military installations, Scoresbysund and Ammassalik are the inhabited areas along the entire east coast of Greenland. There are only three locations along the Scoresbysund Fjord that are inhabited year-round: the village of Scoresbysund and the settlements of Kap Hope and Kap Tobin (Figure 1). The population of all three locations amounts to more than 500 people, mainly Inuit, distributed in approximately 150 households.

Scoresbysund has all the characteristics that, according to Rokkan and Urwin (1983:2) characterize a peripheral society: distance, distinctiveness, dependence, and little control over its fate. The populated area is located at a great distance from other inhabited areas. Contact with other areas of Greenland, not to mention with other areas of the world, is difficult and limited. Passenger traffic to and from the area is possible only by plane, and the fare exceeds the economic means of most residents. Ships can cut through the pack ice for a period of only about one month every summer. Almost all materials and supplies that must be imported are shipped in during this period, and the whole year's production of seal skins are shipped out for further processing.

The society of Scoresbysund has its own, distinctive cultural features, including a dialect that deviates noticeably from the 'standard' dialect spoken on the West Coast. Together with the population in Ammassalik it forms a unique ethnic sub-group within the Greenlandic population. The local lifestyle is greatly influenced by hunting and the rhythm of life is essentially determined by the seasonal cycle, which impacts on the amount and precise variety of game available, as well as on weather and ice conditions.

The local community has a very poorly developed economy. The basis of this economy is limited to hunting (meat for local consumption and skin for export), and public service institutions. The hunting economy continues to have a critical role for most local households. The operation of the public-service system (school, hospital, social security, etc.) within the Scoresbysund society is dependent on an external input of money and personnel. Hunting households are dependent on the international skin market and on the subsidies available from the Greenland Home Rule government. The potential for developing new occupations is very limited, in large part, because of geographical conditions and limited natural resources.[1]

The most important decisions affecting the community are made by individuals and institutions that, for the most part, have very little knowledge of local living conditions or lifestyles. The community itself has few resources with which to limit outside interference and to influence external decision-makers. Furthermore, the society is very vulnerable to external economic, political, and cultural influences. Even seemingly minor changes can have serious consequences for daily life in a society where few collective and individual strategies for survival

---

1. In 1985, ARCO inititiated a survey of the oil potential within an area of 10,000 sq. km in Jameson Land. The activities are still in the beginning phase and it is not certain whether ot not there will be any drilling. The price development of the international oil market and the high costs of operation in this not very accessible area, however, makes it most improbable that an eventual find will be exploited in the foreseeable future.

are available. All these factors contribute to a situation in which the hunting population of Scoresbysund has little control over its own fate.

The Founding and Historical Development of Scoresbysund

The status of the hunting lifestyle and the use of living resources in the Scoresbysund area warrant presentation here as it is against this background that an assessment of the significance of whale hunting for the local population will be made.

Modern-day Scoresbysund was founded in 1925. Although numerous archaeological discoveries show that Inuit have lived in this area for centuries, the region was depopulated sometime around the beginning of the 18th century. The plan to repopulate the area emanated from the wish to find new hunting areas for the rapidly increasing Inuit population in Ammassalik. Furthermore, the Danish government wanted to exercise its sovereignty over the northern part of East Greenland. In 1924 a private committee, the Scoresbysund Committee, provided houses for the future inhabitants and in 1925 about 70 Inuit with hunting equipment and dogs relocated to Scoresbysund from Ammassalik. A trading post was established under the management of a West Greenlandic manager. Soon, a West Greenlandic minister arrived. In the following years a few more families, mostly from Ammassalik, arrived (Robert 1971).

Scoresbysund was chosen for settlement because of an open water area, or polynia, at the mouth of Scoresbysund Fjord. Polynias have traditionally been favorite hunting grounds for Inuit because they attract numerous marine mammals in winter, which in turn attracts polar bears. Moreover, in the summer the polynias are important foraging locations for marine birds. Thus, polynias generally provide excellent hunting potential.

The Scoresbysund area proved to be exceedingly rich in marine mammals. Hunters, in a few years, adapted to the new surroundings, and Scoresbysund quickly became one of the most productive hunting communities in Greenland.

Since the foundation of Scoresbysund in 1925 the local community has undergone a process of modernization. Today, there are a number of modern institutions: a municipal administration and an elected municipal council, a school, a small hospital, a self-service store, a telecommunications center where telephone calls are transmitted *via* satellite to the rest of the world, a local police station, a power plant, a local TV-net and an seniors' home. This development of services has, among other factors, led to a concentration of the population in the small administrative center.

The residents of Scoresbysund live in oil-heated wooden houses, a few of which have running water. Most of the residents own television sets and many have videos. School attendance is compulsory for all children from six to 15 years of age. Many of the young people continue their education outside the municipality, but most return before they complete their education because of difficulties adjusting to the separation.

In contrast to the other hunting districts in Greenland, household consumption in Scoresbysund today includes a large amount of imported products that require a monetary income. As an example, traditional skin clothing has been replaced by manufactured clothing. Furthermore, a rather large amount of European food and stimulants (alcohol, tobacco, and sweets) are sold in the local store. The households also need money to buy hunting equipment and to pay the rent. At the same time, the public-service institutions have created employment for a number of individuals and have thereby made possible a new lifestyle — wage-earning. Some residents have chosen this new alternative, while others have maintained their hunting lifestyle, but only a very few are able to manage without earning a wage during some periods of the year.

Research conducted in Scoresbysund in 1967 revealed that at that time there were 35 hunters in a population of 76 East Greenlandic men over 20 years of age (Robert 1971:96). In 1987 a field survey of local informants revealed that 33 individuals relied on hunting as their primary occupation. An additional 29 individuals hunted on a part-time basis, committing a substantial proportion of their productive work time to hunting activities. Furthermore, 17 individuals hunt in their spare time. Thus, a total of 79 of the men in Scoresbysund spend at least some time hunting (Table 1). In 1987 the East Greenlandic population of men over the age of 20 formed a group of about 130 persons distributed in 100 different households.

Although there have been a number of changes in the life of the community in its 60-year history, modernization has supplemented rather than replaced the traditional hunting culture. The society is still very much characterized as being a hunting society. A hunter is accorded high prestige, and those women who are adept at handling/processing their husband's catch are respected. Hunting is the primary topic of

Table 1.
Active Hunters in Scoresbysund, 1987

Individuals that hunt full-time	33
Individuals that hunt part-time	29
Individuals that hunt in their spare time	17
Total =	79

discussion among the men, and in good hunting seasons, large amounts of hunting products are brought home every day.

Beyond a desire on the part of the Inuit to maintain their traditional hunting culture, the survival of hunting traditions can be explained by the isolation of the community and by the lack of alternative occupations. With the exception of the Thule municipality in North Greenland, Scoresbysund is the only municipality in Greenland where there is no commercial fishing. Many technological innovations have been adopted into the hunting culture (modern hunting weapons and clothing, boats with outboard motors, etc.), and these technological innovations have made hunting more effective. But many traditional elements have also been preserved, the most important of which are described below. But first, the economics of hunting will be described.

The Economic Situation of the Hunting Occupation

As mentioned above, Scoresbysund has in a few years developed into one of the most productive hunting areas in Greenland. The hunting economy experienced a period of relative prosperity after 1955 because of the rising price of seal skin. This development culminated in 1967, when the Royal Greenland Trade Department (KGH) purchased seal skins for a price that was considerably higher than before. By this time, however, the international campaign against the harvest of baby harp seals off Newfoundland had already begun to influence the price of skins for mature seals in Europe. As a result of a series of events over which the society itself had no control, the hunting economy entered a crisis that has now lasted for 20 years. Since 1979, the Greenlandic Home Rule has subsidized the price of seal skins, but this subsidization has merely alleviated rather than prevented the impact of the fall of skin prices on the economy of the households.

In 1967 a good hunter in Scoresbysund had an income from the sale of skins that was approximately twice that of an unskilled worker (Robert 1971:100). In the years immediately before and after this peak year, a good hunter's income was roughly the same as that of an unskilled worker.[2] In 1984 a hunter who wanted an income equivalent to an unskilled worker employed full-time (40 hours a week, 45 weeks a year) had to sell 365 skins of the highest quality. Not included in this estimation are the costs of hunting (bullets, seal nets, fuel, etc.). Assuming these expenses to be 15,000 kroner (roughly 3,000 $ cdn.), 425 skins had to be sold. It is also very important to stress that these skins do not represent the work of one person, but of two persons: usually the hunter and his wife. For comparative purposes it can be mentioned that the most productive hunter sold 310 skins to KGH in 1967, 161 in 1984, and 245 in 1985. Taking into account the fluctuations in hunting production from year to year and the irregular distribution of hunting throughout the year, there is no doubt that 200 to 300 skins represents the maximum annual production for most households. Thus, given the present market, even the most productive hunters are hard-pressed to achieve half the income of an unskilled worker.

Rather than try to produce enough skins to meet economic requirements, most of the households have chosen another strategy: using some of their productive time for wage-earning. A kind of 'occupational pluralism' or mixed economy known for other areas in the Arctic has developed. Local communities have combined subsistence hunting with wage-earning, welfare, and other sources of income in order to be able to maintain a lifestyle that on its own 'no longer pays' (Bowles 1981:91).

On the one hand, the combination of wage-earning and hunting has made it possible for the hunting households to maintain their traditional hunting economy, but on the other hand, the time used for wage-earning has resulted in a decrease in hunting production. This decrease is partly illustrated in Figure 2, which presents an historical summary of the number of seal skins sold to KGH in the period from 1950 to 1986. The

---

2. In 1969 the income from the sale of hunting production was estimated to have covered between 60% and 75% of the east Greeenlandic household's total local purchases from the KNI. In 1984 it covered only 8% to 10% (Larsen 1986:7-8).

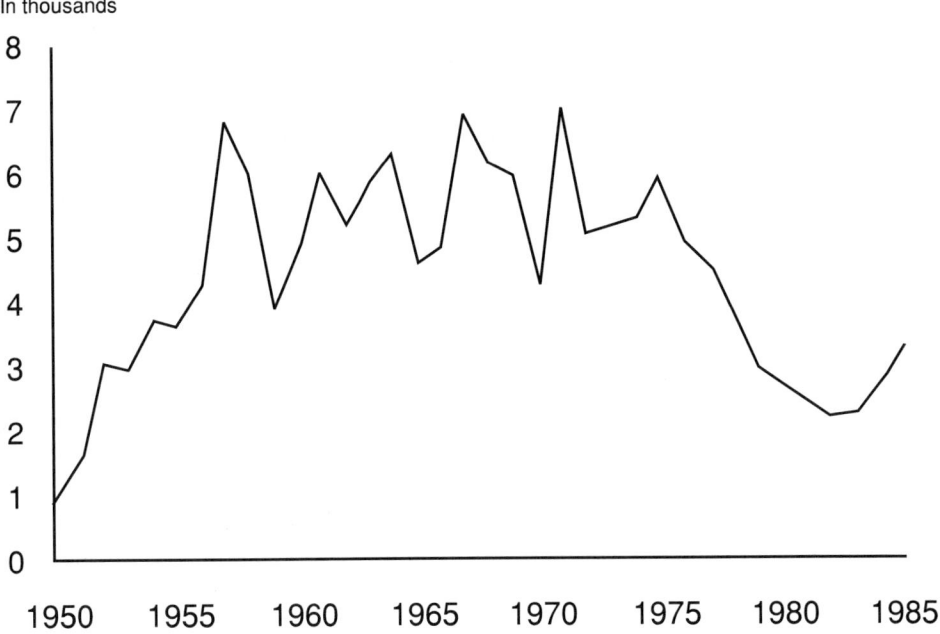

Figure 2. Seal skins sold to KGH, 1950-1985.

number of seal skins sold has declined substantially more than the actual total harvest of seals, because a considerable portion of the seal skins are no longer treated for sale. It is estimated that the untreated proportion of skins has increased from close to 0% at the beginning of the 1970's to over 40% in the middle of the 1980's. The decline in hunting activity has also resulted in a decrease in the supply of meat for private use in the households.

Traditional Elements in the Modern Hunting Culture

The Dog Sled

The dog sled and hunting from the edge of the ice are inextricably linked. The dog sled is used to transport the hunter to the edge of the ice and, no less important, to bring the hunter and his catch home. Furthermore, it is used in polar bear and muskox hunting. A municipal rule forbids the use of 'ski-doos' for hunting. 'Ski-doo' driving is permitted only along the sled tracks between Scoresbysund, Kap Tobin, and Kap Hope. The reason for this prohibition is that the 'ski-doos' scare away the animals. Ownership of a dog team is therefore an absolute requirement for hunting during the eight months that the fjord is covered with ice. The dog sled is furthermore used for transport between Scoresbysund and the two settlements.

A dog team in Scoresbysund typically consists of between 10 and 14 mature dogs. Part of the hunting yield is used for the dog team as dog food. The dogs in the area are generally in good condition, which can be taken as an indication of how much importance is attached to maintaining a good, healthy dog team. In addition to the practical importance, being a good sled driver earns one prestige.

According to Joelle Robert-Lamblin, there were 650 mature dogs in Scoresbysund in 1967 (Robert 1971:92). In 1987 this number had grown to over 730. There were 56 dog teams with eight dogs or more. Among those teams, five belonged to public-service institutions, whereas 51 belonged to private, mostly hunting, households. At any one time, this population of dogs is composed of 150 to 200 puppies under six months of age. Comparing the growth of the number of dogs with the growth of the population in this 20-year period, the number of dogs per east Greenlandic male over 15 years of age has dropped slightly from 6-7 dogs per man to 4-5 dogs per man.

Hunt-Sharing Rules

Rules for sharing the products of the hunt in cases where several hunters have participated in, or at least been present during, the hunt are common among all traditional Inuit societies. Such rules for sharing the products of the hunt are made to foster cooperation in connection with catching larger animals and to prevent conflicts during sharing. In Scoresbysund the local rules for sharing are still strictly followed, and are broadly the same as they were originally in Ammassalik (Sandell 1986a).

Rules for sharing the catch establish the owners of the caught animal and anyone else who has a right to a part of the animal. In regards to smaller seals, the owner is the person who hit the animal first and the seal is not shared with others. For the bigger seals, walrus, polar bear and narwhal, the individual(s) who have shot or even just touched the animal after its capture receives a part of it. For each type of animal, there is a maximum number of persons that can have a share. The shares are given in the order in which the helpers have shot at, or touched the animal, and every share specifies a particular part of the animal. For example, the person who sees the polar bear first is considered the owner of that animal regardless of who might actually kill the bear. The owner is entitled to the skin and to some of the meat. For walrus the rules are similar, with the person who sees the animal first having the right to the head with the valuable tusks.

In regards to the minke whale, there is no owner of the animal. Everyone present during the flensing is free to take a portion of the meat. This kind of collective property of larger whales has earlier been practised everywhere in Greenland. A special kind of collective hunting is the municipal muskox hunt. The hunters that participate are reimbursed for their expenses during the hunt, but they deliver the catch to the municipality, which is then distributed to all the inhabitants of the community.

Food Preferences

Traditionally, meat from marine mammals has been regarded as the only kind of 'real' food. In periods when there was a shortage of meat from marine mammals people starved. Even today the East Greenlandic population in Scoresbysund clearly prefers the meat of seal, whale, and walrus, even though imported European food has gained a footing. The meat is eaten raw (frozen or non-frozen), cooked, dried, or fermented (aged). Blubber is sometimes eaten with the meat. In addition to the meat, most of the entrails of the animal are cooked and eaten as well.

The consumption of meat per inhabitant in Scoresbysund is very high (see below). Numerous investigations in the Arctic show that the traditional Inuit diet provides a more suitable nutritional base than does imported food (Draper *et al.* 1979). Thus, there is a positive correlation between the preferences of the population itself and the recommendations of nutritionists in regards to the composition of the diet appropriate to this area. There is probably also a symbolic dimension in this food preference; by choosing traditional Inuit food one confirms one's identity and membership in a distinct ethnic-cultural group.

The Conception of Time and Attitude Towards Work

In the traditional community, hunting activities were determined by the seasonal cycle, the migrations of game animals, and the weather. Therefore, the rhythm of work was very irregular. In periods of good hunting the men and women worked hard, in periods of bad weather or during the dark period of the year there was less work to be done. During some periods the hunters had to endure watching their quarry for hours in very cold weather or rowing long distances in a kayak in rough water. During other periods hunting needed little effort.

With the introduction of modern institutions and services there followed a new conception of time that was not regulated by changes in hunting conditions or seasons. Wage-earning too requires perseverance, but a different kind of perseverance from that required for hunting. First of all, a constant and regular effort is required. In Scoresbysund a large portion of the population still finds it difficult to adapt to the need for regularity in employment and continues to carry out activities in time with the changing rhythm of the hunting cycle. This is indicated by high turn-over rates and high absenteeism in the work place. For many of those who combine hunting and wage-earning, the former is the activity that continues to sustain one's identity and sense of self, whereas wage-earning is regarded as a necessary evil.

The Altruistic Way of Dealing with Property

The traditional Inuit conception of property demands that one share one's property with others, especially if one has a surplus of an item that others lack. This conception of property is still maintained in Scoresbysund in spite of the introduction of a monetary economy (see sharing of meat below).

Exploitation of Living Resources

The exploitation of resources in the Scoresbysund area revolves almost exclusively around the exploitation of animal resources (Table 2). Altogether 26 species are more or less regularly exploited. These species can be divided into marine mammals, land mammals, birds, and fish and sharks (Born 1973, Sandell and Sandell 1986, Robert 1971). Sea mammals represent the basic resource for this hunting community because the biological production at the mouth of the Scoresbysund Fjord is very high, whereas the vegetation on the adjoining land is scant and limited to but a few months of the year. Berries and herbs provide a vital vitamin supplement to the diet, but quantitatively they constitute an insignificant element in the local exploitation of living resources. The exploitation of the animal resources is described below, with the annual hunting cycle presented in Table 3.

Table 2.
Animal Resources in Scoresbysund.

Marine mammals
Ringed seal (*Pusa hispida*)*
Harp seal (*Pogaphilus groenlandicus*)
Hooded seal (*Cystophora cristata*)
Bearded seal (*Erignathus barbatus*)
Atlantic walrus (*Obedenus rosmarus*)*
Narwhal (*Monodon monoceros*)*
Minke whale (*Balaenoptera acutorostrata*)*

Land mammals
Polar bear (*Thalarctos maritimus*)*
Musk ox (*Ovibos moschatus*)*
Arctic fox (*Alopex logopus logopus/caerulescens*)
Arctic hare (*Lepus arcticus*)

Birds
Brunnich's guillemot (*Urla lomvia*)*
Eider duck (*Somateria mollisima*)*
Black guillemot (*Cepphus grylia*)
Little auk (*Piotus alle*)
Short beaked goose (*Anser fabalis brachyrhynchus*)
Barnacle goose (*Branta leucopsis*)
Red throated diver (*Gavia stellata*)
Kittiwake (*Rissa tridactyia*)
Ptarmigan (*Lagopus mutus*)

Fish and sharks
Char (*Salvelinus alpinus*)*
Sculpin (*Myoxocephalus* sp.)
Greenland halibut (*Reinhardtius hippoglossoides*)
Polar cod (*Boreogadus saida*)
Catfish (*Anarchicas* sp.)
Greenland shark (*Somniosus microcephalus*)

* animals marked with an asterisk are those that are important for the hunting production
(sources: Born 1983, Sandell and Sandell 1986)

Table 3.
Annual Hunting Cycle in the Scoresbysund Area.

	Jan.	Feb.	Mar.	Apr.	May	Jun.	Jul.	Aug.	Sep.	Oct.	Nov.	Dec.
Seal (ice edge)	x	x	x	x	x	x					x	x
Seal (net)	x	x	x	x	x					x	x	x
Seal (on the ice)					x	x						
Seal (open water)							x	x	x	x		
Seal (breathing hole)											x	x
Narwhal					x	x	x	x	x			
Minke whale							x	x				
Muskox								x				x
Polar bear	x	x	x	x	x	x	Protected					
Arctic fox	x	x	x	Protected							x	x
Arctic hare	x	x	x	x	Protected				x	x	x	x
Birds					x	x	x	x	x	x		
Char					x	x			x	x		

Marine Mammals

Four different seals are found in Scoresbysund: the ringed seal, the harp seal, the hooded seal, and the bearded seal. The ringed seal forms a local population that breeds at the head of the fjord, whereas the other three types are more migratory and found in the area usually during the summer period and then only relatively rarely. Another, even rarer, guest is the sand seal. Whereas several thousand ringed seals are caught every year, the annual harvest of all other types combined seldom exceeds 100 seals.

The ringed seal is most intensively hunted during the eight months of winter when the surface of the fjord is covered with solid ice. Younger, non-mature animals from the population in the inner part of the fjord migrate to the entrance of the fjord, where they winter in the polynias. When there is a good solid ice around the polynias, the seals are shot by rifle from the edge of the ice. Another hunting technique is to set up nets under the ice near cracks where the seals appear to breathe. Netting and hunting from the edge of the ice are the most frequently used hunting methods. Under certain conditions, however, additional methods can be employed. For example, when new ice forms in the fall the seals make breathing holes in the ice, which are used for breathing-hole hunting. In the spring, when the seals bask on the ice, they are hunted with rifles equipped with a shooting screen.

In the open-water period from early August to mid-October, the seals are hunted with rifles from ice floes or from a hunting blind on land. A municipal regulation forbids seal hunting from boats with outboard motors. The hunting activity, and consequently the resulting catch of seals, is relatively low during the summer months because hunting conditions are rather poor for two reasons. First, some of the ringed seals migrate away from the entrance of the fjord in August. Second, because the seals have a thinner layer of fat during the summer period, wounded seals often sink before the hunter can reach them.

The walrus is relatively rare in this area. The population does, however, seem to have increased in recent years, with the annual catch ranging from five to 20 animals. Walrus hunting, which occurs during the period from February to mid-June, is conducted by means of high caliber rifles along the edge of the ice from Kap Swainson in the north to Stewart Ø in the south. The walrus is caught both on the ice and in the water.

Scoresbysund seems to be one of the most important summer residences for narwhal on the east coast of Greenland. From February to April they can be observed in small groups along the edge of the ice. From May to July they begin to migrate into the fjords along cracks in the ice. In this period pods of 40 to 50 animals may be observed. During August and September the narwhal stay in the inner part of the fjord. From September to December, when the new ice freezes up, they migrate out of the fjord again. In this period herds of several hundred have been sighted.

In the spring narwhal are hunted by the edge of the ice mainly in the area around Kap Brewster and Stewart Ø, although in some years they are also hunted around Kap Tobin. From August to October they are hunted by boat, both near their summer residence in the inner part of the fjord (Sydkap, Bjørneoerne, Danmarks Ø) and at the entrance of the fjord, and at Stewart Ø. The yearly take amounts to about 10 to 20 animals, which seems rather low considering the number of observed animals. The actual annual catch is limited because there is no use of motor boats at the edge of the ice, and because there is a lot of drift ice at the entrance of the fjord, which makes the hunt both difficult and dangerous.

The minke whale often begins to move in along the edge of the ice at the entrance of the fjord in June. It migrates into the fjord when the ice breaks up, and moves away again when the new ice freezes up in the fall. Normally minke whales may be observed in the area around Kap Tobin and Kap Swainson either singly or groups of two or three. Minke whale hunting takes place at the entrance of the fjord in August and in the beginning of September. When the entrance of the fjord is free of ice, a large number of boats (15 to 20) with outboard motors approach the animals, which are shot with high caliber rifles and are harpooned with one or two harpoons equipped with floats. The animals are normally transported to Kap Tobin, where they are skinned and butchered. The yearly take during the period that the minke whale has been hunted has fluctuated between 1 and 11. This considerable variation in the annual catch results from varying amounts of drift ice, which in some years makes minke whale hunting very difficult.

Land Mammals

The polar bear is perhaps the most unpredictable of the animals hunted in the Scoresbysund area. Polar bears may be found anywhere in the area inside the fjord and along the Blosseville Kyst and the Liverpool Land Kyst during all periods of the year. Occasionally bears are shot close to the inhabited areas or by seal hunters at the edge of the ice. But there is also a polar bear hunting season from February to June when some hunters, mostly younger, unmarried men, go on a polar-bear hunt for a period of up to one month. These hunting trips, which are made on dog sleds, normally cover a territory of more than several hundred kilometers either north in the direction of Kong Oscars Fjord and Trail Ø, south along Blosseville Kyst or into the inner part of the fjord to Danmarks Ø and Nordvest Fjord. The relationship between the time and effort expended on these hunting expeditions and the number of polar bears actually caught is extremely uncertain, which is why family providers normally prefer to hunt more predictable species. For the younger hunters, however, it is not only the skin and the meat that are desirable, but also the prestige accorded a good bear hunter. Approximately 30 to 60 polar bears are shot each year in the Scoresbysund area.

The muskox was an animal unknown to the new arrivals from Ammassalik when they moved into the area in 1925. They soon learned, however, to appreciate its delicate meat. Right from the beginning of the colonization of Scoresbysund there has been a policy of allowing muskox hunting only in case of an acute shortage of meat. On this basis, a tradition of a municipal hunt of muskoxen on Jameson Land, with a subsequent distribution of meat to the inhabitants, has developed. In recent years there have been two joint hunts each year, one in September and one in November. Most hunts target an average of one muskox for every 15 inhabitants. In addition to the official municipal hunts, there is some illegal muskox hunting. Taking this into account, the total amount of animals taken is judged to be about 400 to 500. This number seems more or less to be the maximum sustainable yield. Hunting takes place both in winter, when the meat is transported home by sled, and in the summer, when the meat is transported home by boats with outboard motors. Muskox hunting, which is less dependent on luck than is polar-bear hunting, seems to have increased since the decline of seal skin prices which has adversely affected the hunting economy.

In the decade preceding World War II the hunting of the Arctic fox was the most important source of cash income for hunters in Scoresbysund. However, fox hunting declined after 1960, when the price of a seal skin reached the price of a fox skin — for a long period a seal skin commanded only about 1/10 the market value of a fox skin. Moreover, the fox is taken only for the value of its skin. Since the seal, in addition to its skin, also contributes a valuable supply of meat and blubber, it was clearly more rational to concentrate on seal hunting. Some hunters still catch foxes, but only in limited amounts.

There is a substantial population of Arctic hares in the area, but there is no systematic or routine hunting of hares. They are shot, if they are seen by chance, in the course of other hunting activities.

Birds

In May the area around Scoresbysund is invaded by hosts of marine birds that breed in the area around the entrance to the fjord (Kap Brewster), along the south coast of the fjord (Volquart Boons Kyst), and along the north coast (Liverpool Lands Kyst, Raffles Ø, Rathbone Ø, Kap Høegh). The most common of the marine birds are the guillemot and the little auk. The birds forage in the polynias and are hunted from the edge of the ice in connection with seal hunting in May and June. In the same period some migrating geese are occasionally shot. After the break-up of the ice, the guillemot is hunted from boat or from ice rafts in connection with seal hunting. Young glaucous gulls are hunted at the end of August and in the beginning of September. In June, eggs are collected from the bird colonies. Birds, which contribute only a minor proportion of meat production, are most commonly hunted when other hunting activities are taking place. However, the meat of the birds adds welcome variation to the diet and the collection of eggs functions as a social activity that gives the families an opportunity to spend some time together in the camps.

Fish and Sharks

The most important sources of food for marine mammals and marine birds that forage at the entrance of the fjord are pelagian crayfish and polar cod. Neither of these species are attractive for fishing (polar cod is, however, caught in insignificant quantities from the ice), and the area is in general ill-suited to fishing. Occasionally catfish, Arctic halibut and sculpin are caught. As in the case of the marine birds, fish provide a little variety to the diet. In the spring Arctic char are caught through the ice on the lakes near the inhabited areas, and in August they are caught by nets in various areas in Hurry Fjord and in the area around Sydkap. The yield is maximally 2000 to 3000 char a year. Although the catch can hardly be considered to cover the costs of fuel for the motor boats, the fishing activity in itself has an important social and recreational function. In the spring Greenland sharks are fished from the ice. The non edible meat is dried and used as dog food. Approximately 50 to 100 sharks are taken each year.

The Use of the Products of the Hunt

Figure 3 shows the flow of resources in a hunting household. There are three raw material groupings: skin, edible organic tissue (meat, blubber, and entrails), and ivory, bones, etc.

Skin

Formerly, a major proportion of skins taken in households was used in the manufacture of clothing and traditional skin boats (kayak and *umiak*, or traditional woman's boat). However, now the skins are very seldom used for these purposes. This change has occurred partly because inexpensive manufactured goods have replaced traditional skin clothing and because the kayak is rarely used for hunting anymore. Similarly, the *umiak* and skin tents have also been replaced by modern equivalents.

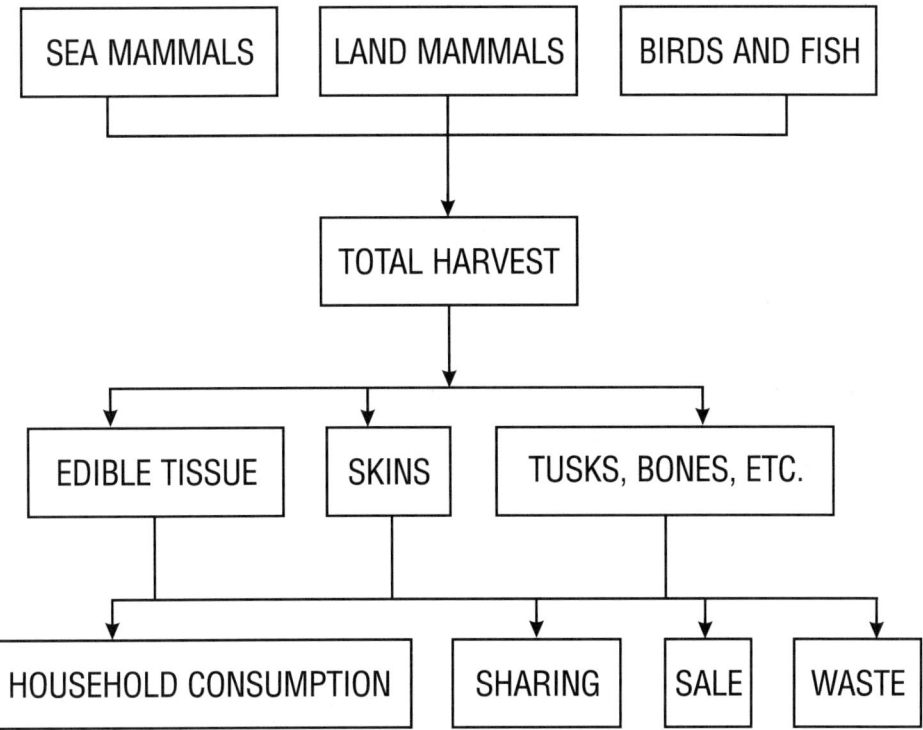

Figure 3. Flow of resources in a hunting household.

Seal skin

Most seal skins are cleaned of blubber, scraped, washed, stretched, and dried. Processing skins is traditionally considered to be women's work. When the skins are ready, they are sold to the local, home-rule owned store (KNI, formerly KGH), and are priced according to size and quality. Only rarely are skins sold privately. Skins from seals that are shot during the molting period (May to June) are normally used as dog food since such skins have little commercial value. Furthermore, as mentioned above, the serious decline in skin prices that began in the middle of the 1970's has so eroded the market that a growing percentage of the skins obtained during the other periods of the year are not sold either. The time required to treat the skins is simply too great relative to the economic yield, and in many households women have chosen to join the paid workforce rather than try to rely on this marginal source of income. A rough estimate indicates that 40% of the skins are not sold, but rather are fed to dogs. Sometimes a hunter gives skins away to relatives if he himself has no wife who can treat the skins.

Polar bear skin

Almost every polar bear skin is sold because these skins are very valuable. Some of the skins are sold to Danish artisans and employees because the hunters can get better prices through such private sales than by selling them to the KNI store. Skins not sold privately are sold to KNI.

Muskox skin

Some of the muskox skins are used for sleeping bags or as covers for the dog sleds. The rest of them are sold privately because the KNI store does not purchase these skins.

Fox skin

Fox are hunted only for their skin which are sold to KNI.

Edible Tissue

In contrast to skins, which are for the most part sold, only a very small portion of the household production of meat, blubber, and entrails is sold. Most of this edible tissue is used in local subsistence. That portion that is consumed within the household is used both for human food and dog food. Some items (e.g., shark meat and certain entrails) are exclusively used as dog food, whereas others (e.g., polar bear and muskox meat) are exclusively used as human food. This does not include meat used to feed dogs during the transport of the catch from the hunting site to the village.

During those periods when the seal hunting is good, the dogs are primarily fed seal meat and blubber. Dogs are also fed whale meat, walrus meat, and imported European dog food. In those households where there is a shortage of meat, the need for human food is met first. Dogs, in such cases, are given the leftovers of the household's own produce and bought food, if the household is not so lucky as to get some dog food from neighbours or relatives who have some spare meat.

Households that have a surplus of meat will normally give it away to neighbours or relatives. Smaller amounts of meat are sold to various institutions in town (e.g., seniors' home, hospital, municipal council, Greenland Technical Organization), where the meat is used as human food or dog food for the institution's sled teams. Some meat is sold to Danish households in the town of Scoresbysund. It is more common now to request payment from more distant relatives for some of the more desirable of the products of the hunt.

There is no export of meat out of the district. Until 1963, the KGH bought blubber, which was then exported to Europe, where it was sold primarily to the margarine industry. Today, much of the blubber is wasted, since only a part of the total production of blubber is consumed by humans and dogs, and since there is no alternative means of using surplus blubber in the local economy (traditionally blubber was used for cooking and heating the houses). Whereas new conditions have led to a certain amount of blubber wastage, wasting of meat is almost unknown. When there is a surplus of meat, the excess is frozen or dried. Meat that is no longer suitable for human consumption is usually used for dog food.

Tusks, Bones, etc.

Tusks from walrus and narwhal were formerly used for the production of tools, etc. Today, the tusks are sold to KNI or to interested individuals. Bones and hoofs are rarely used nowadays.

Local Production of Food, 1985

Local production of food in Scoresbysund for 1985 and its distribution as human food, dog food, and waste can be estimated within limits. Table 4 shows how many animals of each species were taken. These numbers are based on official hunting statistics (Anon. 1987), supplemented with non-official sources, and are subject to a certain amount of uncertainty. The numbers should therefore be viewed as a 'best guess.' The number of animals caught is multiplied by the estimated weight of the edible tissue (meat, blubber, and entrails) per animal in each category, with the exception of the entrails of minke whales (Sandell 1986b). As indicated in Table 4, about 200 tons of edible tissue were produced in Scoresbysund in 1985. It is important to note that the harvest of this year must be regarded somewhat better than the years immediately before.

The distribution of this edible matter as human food, dog food, and waste is calculated in Table 5. The calculation takes into account the following considerations, first the need for dog food is estimated on the basis of information concerning the local practice of feeding dogs (Sandell 1986b). The total need is estimated to be 227,213 kg of edible tissue. However, part of this food need is covered by imported dog food,

Table 4.
The Production of Edible Tissue in Scoresbysund, 1985.

Type of animal	Harvest (in numbers)	Edible Tissue (kg)
Seals	4,778	132,690
Minke whales	10	20,000
Narwhal	28	10,640
Walrus	22	10,956
Polar bear	23	3,680
Muskox	400	22,000
Fish, birds	-	20,000
Total =		**219,966**

it is estimated that 25,000 kg of imported dog food was purchased from the local store in 1985, approximating 103,463 kg of edible tissue. The amount of imported dog food is subtracted from the total need. On this basis it is estimated that 123,463 kg, which is 56% of the total production of edible tissue, is used for dog food. It is further estimated that 10%, or approximately 21,996 kg, of the edible tissue has been wasted. The remaining 74,507 kg, corresponding to 34% of the total production, is used for human consumption. Once more, it is necessary to stress that the numbers are subject to some uncertainty.

From Table 5, it is possible to calculate the average use of locally produced food per inhabitant each day in 1985 to be 436g (as of July 1, 1987, 468 persons born in Greenland resided in Scoresbysund). Inasmuch as this figure includes animal food, it is rather high. But the figure is significantly below the estimated total optimal consumption of meat per inhabitant of 1 kg per day. The rest of the need for food is covered by imported food. It should also be noted that the high use of locally produced food has been possible only because 54% of the need for dog food is covered by imported products. The heavy reliance on imported dog food illustrates that the catch of traditional game animals is no longer sufficient to cover the total need for these products within the society.

Table 5.
The Uses of Edible Tissue as Food and Waste.

Use	Edible Tissue (kg)	Percentage
Human food	74,507	34 %
Dog food	123,463	56 %
Waste	21,996	10 %
Total =	**219,966**	**100 %**

The direct connection between this insufficiency and the critical situation facing hunting as an occupation today is discussed below.

The Importance of Minke Whale Hunting in the Economy

The importance of minke whale hunting for the local community is best evaluated within the overall background presented above, especially with respect to the economic situation of the hunting population and its use of the renewable resources in the Scoresbysund area. Since the foundation of Scoresbysund almost every year a great number of minke whales have been seen in the mouth of the fjord during the open-water period. But it was not until 1980 that the first two minke whales were caught. Until that time there was a lack of means of transportation appropriate to this kind of whale hunting. In the beginning of the 1970's the first boats with outboard motors were introduced into the area. Within a few years, almost every hunter owned a motor boat. This technological innovation made the cooperative hunting of minke whales possible, as 15 to 20 boats are now able to participate collectively in the hunt.

The hunting of minke whales can therefore be regarded as the most recent adaptation of the Scoresbysund to an ecological niche they settled into in 1925. In this connection, it must be stressed that Inuit society has never been static. Rather, it has always been able to adapt to changes in the resource base and to integrate new methods and equipment into the hunting culture.

Table 4 indicates that the hunting of minke whale in 1985 contributed almost 10% of the total production of edible tissue. This amount can be interpreted as a lot or a little, depending on one's point of view. The minke whale belongs to a species that provides an important part of the production of edible tissue in the area, but seal hunting is still the activity that provides most of the food for the population.

The importance of the hunting of the minke whale is, however, greater than the production figures might suggest because it takes place during a period of the year when only a few seals are caught. Therefore, the hunting of the minke whale has helped to stabilize the resource base for, and fill an important gap in, the hunting economy. This contribution to the general health of hunting as an occupation is especially important during the current period when the hunting culture is threatened by external factors and when there are clear indications of an overall decline in the subsistence production.

Conclusion

The subsistence economy of Scoresbysund has developed into a mixed economy. This development is reflected in the human diet, which can also be described as a mixed diet. Compared to other areas of Greenland it has been possible to maintain a relatively high consumption of locally obtained meat per inhabitant, as there are still many hunters in the area. In recent years, however, hunting as an occupation has been subjected to heavy pressure because of the low price of seal skins. Nevertheless, the following conclusions can be drawn:

- Meat from marine mammals is the cornerstone of the hunting culture, as it is regarded as the only kind of 'real' food. Furthermore, marine mammals make dog teams possible, which are a fundamental component of the hunting economy.
- A comparison of the production of meat in 1985 with the optimal need for meat shows that there is a deficit. This deficit of meat can be explained, in part, by the fact that hunters are increasingly being forced to turn to wage labour positions since the decline in the price of seal skins.
- The falling price of seal skins is threatening the production of an important source of cash and the production of a vital food source.

- Any interferences or external influences that limit the production of meat also undermine the basis of the hunting culture.
- Even though the catch of minke whales contributes a small portion of the total production of meat, it is nonetheless of considerable importance, because this whaling takes place at a time of the year when sealing productivity is minimal.

Bibliography

Anonymous. 1987. *Sammendrag af Grønlands fangstlister m.v,* Nuuk.
Born, E.W. 1983. *Havpattedyr og havfugle i Scoresby Sund:* fangst og forekomst, København
Bowels, R.T. 1981. *Social Impact Assessment in Small Communities.* Toronto 1981.
Draper, H.H. *et al.* 1979. Report of the Nutrition Panel for the Aboriginal/Subsistence Whaling Panel Meetings, International Whaling Commission, Seattle.
Larsen, F.B. 1986. *Pengeindkomster i et fanger-lønarbejdersamfund*, Nuuk.
Robert, J. 1971. *Les ammassalimiut émigrés au Scoresbysund*, Paris.
Rokkan, S. and D.W. Urwin. 1983 *Economy, Territory, Identity: Politics of European Peripheries,* London.
Sandell, B. 1986a. *Om fangstdeling i Scoresbysund Distrikt,* Gudhjem.
Sandell, B. 1986b. *Beregning af hundefoderforbrug i Scoresbysund,* Gudhjem.
Sandell, H. and B. Sandell. 1986. Kap Hope: A Settlement and its Resources. *Arctic Anthropology* 23(1-2): 281-298.

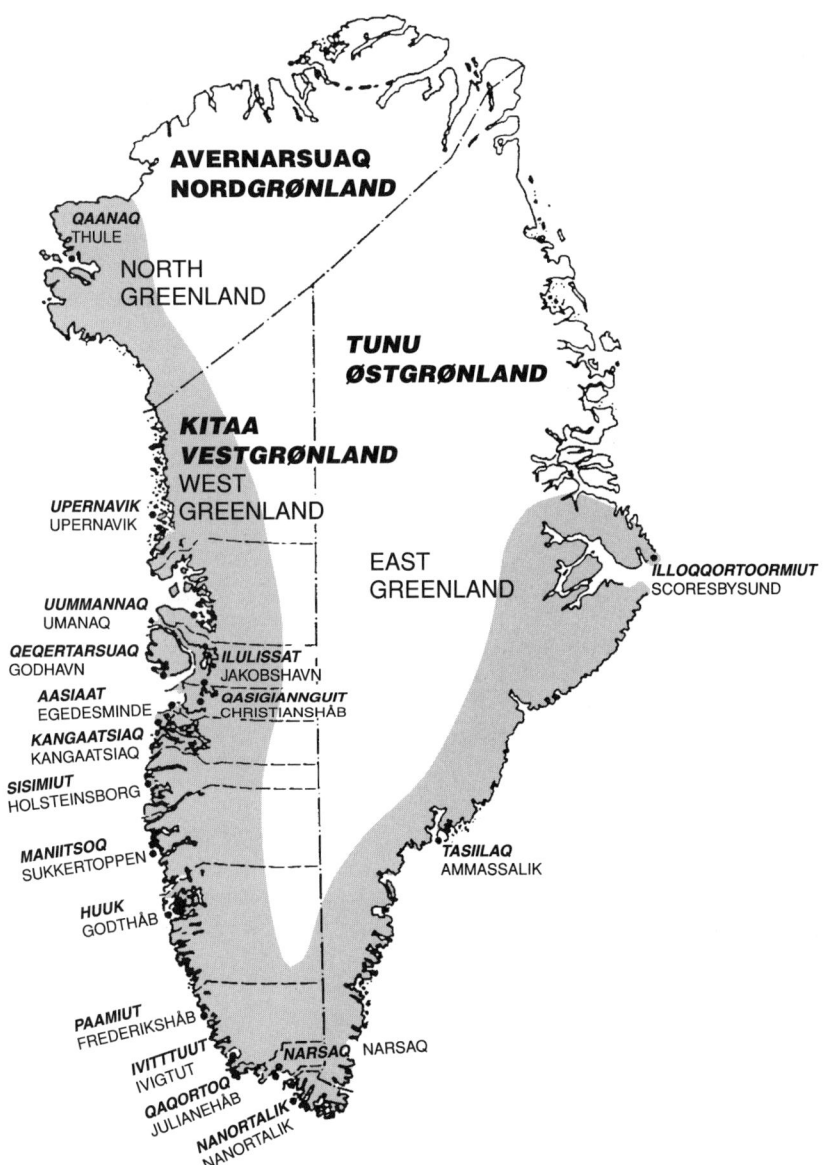

Community-based Whaling in Greenland.

Greenland Subsistence Hunting[1]

Greenland Home Rule Government

Janne Jervin
(Project Coordinator)
with contributions by
Jens Dahl, Peder Helms, and Robert Pedersen
1989

Introduction

The International Whaling Commission (IWC) adopted a motion to freeze whaling from 1986 through 1990, when the resolution will be reviewed. The ban affects commercial whaling, but not traditional whaling by indigenous peoples. The reason for the latter exemption is that

---

1. Editor's note: Not quoted in extenso. Paper has been extensively reorganized and shortened; sections dealing with dietary and socioeconomic projects relevant to the subsistence issue, environmental protection and 'wildlife management' programs, and registration of natural resources have been left out. Those wishing to consult the latter are urged to contact the IWC or the Greenland Home Rule government.

traditional whaling is perceived as being fundamentally different from commercial whaling.

Most people are familiar with the popular visual images of high-speed whaling vessels and huge factory ships with several large whales in tow. Catching and processing of the commodity are done with the use of high technology and profit is the incentive that drives the operation. A limited number of companies operate well-organized whaling businesses whose success is tightly bound with the sale and export of products and by-products.

What Actually is Traditional Whaling?

Aboriginal subsistence whaling means whaling for purposes of local consumption carried out by, or on behalf of, aboriginal, indigenous, or native peoples who share strong community, familial, social and cultural ties related to a continuing traditional dependence on whaling and on the use of whales. Local aboriginal consumption means the traditional uses of whale products by local aboriginal, indigenous, or native communities in meeting their nutritional, subsistence, and cultural requirements. The term includes trade in items which are by-products of subsistence catches. Subsistence catches are catches of whale by aboriginal subsistence whaling operations (from Kapel and Petersen, *Subsistence Hunting: The Greenland Case* 1979).

This somewhat lengthy definition of traditional whaling reflects the complexity of the concept and hints at the difficulty of associating a single definition with subsistence whaling. The definition must, by necessity, include the historical, cultural, technological, nutritional, and economic aspects of subsistence whaling — aspects which may be expressed in highly different ways, depending on the particular field under consideration. To deal with subsistence whaling, one must include all these aspects in one's conclusions. The difficulty of fitting the various elements into an evaluation of subsistence whaling can be appreciated by studying the ancient traditions of whaling carried out by Arctic peoples.

An evaluation of this sort will frequently, and logically, focus on the hunting methods used. How sophisticated must hunting gear become for it to be no longer classified as traditional? It must be understood that hunting technology has evolved steadily from time immemorial towards methods and gear which are easier and more efficient to use.

Hunting equipment has evolved steadily through the nearly 5000 year history of Inuit culture. Today, many people feel uneasy about the suffering of a whale when caught by traditional methods. This attitude poses a dilemma.

The 'time to death' of whales will definitely be reduced with the use of high-tech hunting gear (hot-grenade). But the introduction of such weapons is itself the source of a new problem; it clashes with traditional whaling. By responding to the call for more humane killing methods in subsistence whaling, whale hunters are immediately confronted with the issues stemming from the popular misconception that subsistence whaling should not be 'high tech.' This explains the hunters' reluctance to switch to the penthrite grenade. Experiments show that the penthrite grenade is eminently suited for whaling. The minke whale dies instantly; the time to death reduced to seconds. But the penthrite grenade is expensive and can at present be used only by vessels equipped with harpoon guns. Hunters operating from small boats cannot benefit from this development. If subsistence hunters are to switch to weapons that match the penthrite grenade in sophistication and efficiency, extensive travel, study, and information campaigns will be required, all of which are beyond the means of Greenlandic hunters and for which they are reluctant to seek financial assistance. It is all quite alien to them.

On the other side are the leaders of the hunters' own organizations, KNAPK, and the Greenland Home Rule government that want the hunters to understand the necessity of accepting what they consider to be an irreversible trend. The higher echelons in Greenland recognize the need for progress in this field, but they also want the Greenlandic hunter to contribute his own understanding and political will to the cause. This is why things have moved more slowly than originally scheduled in the organization of fishermen and hunters in terms of preparing for the general introduction of the penthrite grenade. However, given the recent, political consent in the organization, initiatives taken by KNAPK the coming year are likely to produce substantial progress.

The main purpose of this document is twofold. First, to comply with requests for updated information about Greenland in general. Second, to give the world around us an impression of the important commitment undertaken by the Greenland Home Rule government to research and manage the natural amenities and living resources of the country. The intention, of course, is to establish a balanced utilization of Greenland's natural resources and to provide the data required to set up a long-term and prudent strategy to govern utilization of the living resources as part of circumpolar cooperation in this area.

Toward this end, we have deemed that a general, factual description of Greenlandic society would prove useful. Included in the text itself are only such tables and figures which are considered essential for the sake of clarity. This approach applies to all sections.

It is the wish of the Greenland Home Rule government in the following sections to furnish the reader with an idea about subsistence hunting in Greenland from the first hunting cultures, up until today. According to existing statistics, a modest percentage of the population is engaged in hunting. Reality presents a different picture, however, since fishing and hunting linked to subsistence, outside the germane hunting districts of North and East Greenland, constitute a crucial, viable way of life which often escapes statistical treatment.

The various aspects of large whale hunting throughout the centuries are dealt with extensively. The dietary and nutritional aspects, which was one of the reasons underlying an initiative launched by the Danish authorities in the 1920's to resume the hunting of large whales, are also described. The aim of the initiative was to avoid hunger and malnutrition during a period in which a drop in the seal population, poor weather, heavy ice, and epidemics ravaged the country. This type of hunting was discontinued in 1959, but since this period represents an important part of the full 'hunting history' in terms of health and nutrition, its description is warranted.

At the insistence of the Greenland Committee of 1948, a thorough dietary study was implemented. Published in 1955, several studies of diet and nutrition were to follow. But changes keep coming, and the Greenlandic Society has, particularly since the 1960's, seen a dramatic development in diet, health, and nutrition which continues unabated. The data regarding diet and nutrition have not been adequately updated since the 1970's. Collection of data on the socioeconomic aspects of hunting and small-scale fishing has been limited to a number of separate studies.

Jens Dahl describes the cultural role of hunting and subsistence in Greenland. Concerns are raised regarding problems of even tackling the job of defining subsistence hunting. Dahl has made a valuable contribution based on a multi-case study from the Saqqaq settlement in the Disko Bay. He discusses both general and key problems associated with definitions of subsistence fishing and hunting.

Traditional and present distribution channels in subsistence hunting in Greenland are covered by Robert Petersen. He illustrates why Greenlanders today must, in the absence of equal conditions for bartering and sharing the catch, have recourse to another kind of asset — money. Whether one chooses to reciprocate in kind or through means of payment most common today, i.e., money, what happens is more or less the same. In any event, this is the only realistic way of keeping the game sharing channels of distribution open.

Over the years, the IWC has been presented with copious material on subsistence whaling in Greenland. Most of these have been presented by Danish/Greenlandic delegations and many appear in edited or slightly altered forms in this volume.

General Facts About Greenland

Introduction

The objective of this section is partly to give a factual description of Greenland, its geography, ice conditions, inhabitants, business sectors and administration, and partly to give an outline of its historical development from the beginning of historical times where legends meet historiography, to the Greenland of today. Knowledge of the above is essential to understand the present living conditions of Greenlanders.

Geography

Greenland, the world's largest island, is part of North America and is separated from Canada by the Nares Strait, which at its narrowest point is only 26 km wide. This geographical location has affected the historical development of Greenland, since its fauna and first settlers came from the west. Spitzbergen is approximately 500 km to the east, while the Denmark Strait, separating Greenland and Iceland, is about 275 km wide.

The northernmost point of Greenland, Cape Morris Jesup at 83°N, is only 740 km from the geographical North Pole, while Greenland's southernmost point, Cape Farewell, situated on an island separated from the mainland by the Prins Christian Strait at 60°N is located on the same latitude as central Scandinavia. Greenland extends 2,670 km from north to south and 1,050 km from east to west. The total area is approximately 2,415,000 km^2, of which 385,000 km^2 is not covered by ice cap. By comparison, Greenland is almost 10 times the size of the United Kingdom. The ice cap, which at some places is more than 3,000 m thick, covers some 1,834,000 km^2 and contains an estimated 2.7 million cubic kilometers of fresh water ice, or close to 10% of the total freshwater supply of the world.

Climate

Twenty-three degrees of latitude separate the northernmost and southernmost points of Greenland. Owing to this vast expanse, the climate varies widely between north and south Greenland. On account of the ice cap, the climate, even in South Greenland, must be described as Arctic, as the annual average temperature is around or below freezing point, and the average temperature does not exceed 10°C even in the warmest month of the year. The special character of the climate on the ice cap derives from the fact that snow and ice surfaces always will remain cold. The lowest temperature registered in Greenland, about -70°C was measured on the northern part of the ice cap.

A distinctive feature of the ice cap's boarder zones is the frequent, very violent storms with drifting snow, which also reach the country in front of the ice cap. On the ice cap, annual precipitation averages more than one meter of snow, or 34-35 cm of rain fall. The south (e.g., Cape Farewell), averages over 200 cm per year, while the north (e.g., Peary Land) averages less than 10 cm.

The climate of the southern coastal areas of Greenland is scarcely colder than several places in central North America, but the winter is much longer and the summer shorter and cooler. However, the climate of the outer coastal areas may differ from that of the far-inland areas, where the climate may be more continental.

It is characteristic of the Arctic that the soil at a certain depth is frozen permanently; only the upper layers have time to thaw during the short summer. This phenomenon is called 'permafrost.' It causes considerable problems for building and construction as the upper layers become very unstable, making the laying of foundations for buildings difficult.

Almost all of the ice-free coastal area is mountainous; the highest elevations are to be found on the east coast at about 4,000 m. Large numbers of reefs as well as islands of varying size where many of the settlements are situated are found along the coast. The coastline, which is more than 40,000 km long, is also characterized by a large number of bays and inlets, some of which are among the largest in the world.

Vegetation

Trees grow very slowly in the Arctic. There are no forests in Greenland, and with rare exception, plants cannot be cultivated at a profit. However, wood-like scrub is found at several places and the warmest valleys in south Greenland boast 20 feet high birches. Here can also be found the Greenland mountain ash, and various species of willow. Under the trees an abundant flora of fern and herbs dominates. Despite its northern location, Greenland has a surprisingly rich flora. The most widely distributed plant community in Greenland is mountain heather, which in North Greenland is dominated by the cassiope, and in south Greenland, primarily by crowberry and dwarf birch. Mountain aven is widespread, and as soon as the ice has melted, the purple saxifrage begins to bloom. Marsh and meadows with cotton-grass and various other grasses are widely distributed and may form thick layers of peat, which, in former times, was used by Inuit for the construction of their huts. In some settlements, inhabited huts are found where at least one of the walls is built of stone and peat. In other places, these houses have been preserved or used as storeroom/outhouses.

In most cases, Greenland's soil is fertile, and where the ground is fertilized naturally, the vegetation can become astonishingly rich. Here and there, an especially lush green colour and richness of the grass indicate the existence of old Inuit house remains, where refuse from humans and game has enriched the soil. The same rich vegetation can be found at the base of bird cliffs. Fern and harebell grow in fissures, and thick and luxuriant moss vegetation can be found where the water seeps down cliff walls. Stones and dry areas are dominated by various kinds of lichen. On slopes facing south, especially in West Greenland, grow Greenland orchids, and angelica, the tallest plant in Greenland, the stalks of which are edible and replace the rhubarb of more southerly latitudes.

Ice

The polar ice off East Greenland drifts southward until Cape Farewell, where the Gulf Stream takes it northward. It melts eventually and seldom reaches further than the Nuuk/Godthåb area. Periodically, the polar drift ice is a hindrance to the fisheries. Under normal winter circumstances, in the area of Paamiut/Frederikshåb, up to and including Sisimiut/Holsteinsborg, local ice is only of little consequence to navigation along the coast. The area from the Disko Bay to Upernavik is not navigable during the months from December to April/May. Nor is Thule navigable in

October and November. Ammassalik is only navigable in July to October and Ittoqqortoormiit/Scoresbysund in July to September; both can only be visited by ships with reinforced hulls.

Population

The indigenous population of Greenland are the Eskimos or the Inuit. Culture and language bind the Inuit of Greenland to Inuit in Canada, Alaska, and Siberia. As of 1 January 1988, Greenland had 54,524 inhabitants, ca. 4/5 of the population being native Greenlanders. The last fifth is made up of Danes and people of other nationalities. The population is distributed throughout many small settlements and a few larger towns. Out of a total of 120 settlements and towns, 65 are inhabited by less than 100 people. There are about 12 towns with more than 1,000 inhabitants. Today, about 3/4 of Greenland's population live in these towns.

Language

Eskimo and Greenlandic

Greenlandic is the eastern branch of the Eskimo (Inuit) language spoken across the Arctic from the eastern tip of Siberia to east Greenland. Aleut is the only tongue remotely related to Eskimo, which is an agglutinative language with no relation to the Indo-european family of languages. Southern Alaska and Siberia form their own language group whereas, linguistically, northern Alaska is related to Canada and Greenland. Roughly, Greenlandic can be divided into East Greenlandic (Tunu), West Greenlandic (Kitaa), and Polar Eskimo (Avanersuaq), in conformity with geographical divisions.

Naturally, association with the Europeans has caused some European expressions to be absorbed by Greenlandic. The oldest of these words have now been completely adjusted to the Greenlandic form of speech. The colonists never tried to force their own European language onto the population of Greenland. The objective of this policy was no doubt to preserve that part of the local culture which did not collide too much with the then Christian philosophy of life. At a later time, attempts were made to create a more systematic Eskimo grammar and to establish orthographic rules.

In the middle of the 19th century the missionary and teacher, Samuel Kleinschmidt, initiated an analysis of the language. This analysis has ever since been regarded as outstanding, and he also prepared rules for standard orthography which were not revised until the start of the 1970's. In 1857, Inspector Rink began, as the first, to print Greenlandic on a large scale. Shortly after this significant achievement, illiteracy dropped swiftly among the indigenous population. In 1861, the esteemed old newspaper in Greenlandic *Atuagagdliutit* was established. This newspaper has been published ever since and is among the oldest in the world.

In the schools only Greenlandic was taught and gradually more and more subjects were covered by books in Greenlandic. However, school books were not the first books to be printed in Greenlandic. Long before these, missionaries published religious literature in Greenlandic. Along with the generally accepted rules governing standard orthography, other kinds of literature emerged, and several Greenlandic poets and novelists had their work published in the first half of the 20th century. By then, the Greenlanders were able to read and write their own language fully. A UNESCO publication estimates that in the 1960's, only 1-2% of the Greenlandic population was illiterate. During the same period, illiteracy likewise accounted for 1-2% on the Faroe Islands and in Denmark, compared to 10-15% in Alaska and in the Caribbean, 5-10% in Israel, and 40-45% in the Mongolean Peoples' Republic.

Trade and Industry

Due to Greenland's geography and environment, commercial activities are rather limited. The population has lived and still does on subsistence practices, and has primarily been forced to earn its living, directly or indirectly, by hunting and fishing. The difficulties caused by natural conditions and the country's remote location are great. These circumstances call for extraordinary aptitudes and particularly vigorous efforts in comparison to what is needed under more favourable circumstances if an acceptable standard of living is to be attained, let alone maintained.

The indigenous Greenlandic way of life permitted a standard of living which was modest compared to Denmark and other countries at a similar development level. This was mainly due to the fact that during colonial times (i.e., up to the years after the Second World War) greater emphasis was placed on safeguarding the indigenous population from external influences, rather than on increasing material welfare. Consequently, the standard of living was generally determined by the traditional way of hunting and fishing. Almost all Greenlanders remained hunters until climatic and other changes forced more and more to take up fishing. This pattern gradually changed after the Greenland Commission of 1948 recommended an improvement to the standard of living and the modernization of the Greenlandic community.

In principle, a rather well-defined distinction was established between high-level capital transfer public contributions on the one hand, and on the other, private income (income of the population being tied to its own level of activity, which was predominantly fishing or hunting). During the 1950's the standard of living, and especially the standard of health, improved. The mortality rate fell, and the birth figures rose.

A glance at developments in the distribution of labour on the main occupation can give an impression of the impact of the development policy launched in 1950. At the start, about 55% of the labour force was employed in the primary industries (hunting, fishing, and sheep-farming), about 13% in public institutions, liberal professions and services, while a little less than 27% was employed in production (including fish processing), construction activities, commerce, and transport. In 1965, people engaged in primary industries fell to 27.5%. At the same time, the total labour force had doubled. In 1970, primary industries accounted for 19.1% of the work force.

Hunting as an occupation is performed throughout Greenland. In east Greenland, as well as the Upernavik and Uummannaq municipalities, inhabited by altogether approximately 1/5 of Greenland's population, hunting is the primary means of subsistence. Disko Bay is also an important hunting area, and this occupation is the principal source of income for the outlying settlements. Subsistence hunting of terrestrial and marine mammals, birds and fish is essential in all Greenland, including the towns and settlements of central and southern Greenland, even in the Nuuk/Godthåb area. In Central and Southwest Greenland, where the number of people engaged in hunting is relatively smaller than in the traditional hunting areas, subsistence hunters and fishermen are as much dependent on their occupation as people in the outlying districts. The socioeconomic and cultural aspects concerning hunting and fishing with be elucidated in future sections.

Hunting Licenses

The municipal lists of valid hunting licenses show the number of people entitled to professional hunting in Greenland (cf. Hjemmestyrets bekendtgørelse nr. 1. af 19. Januar 1981 om udstedelse af fiskeri — og fangstbeviser, Appendix, p. 201). Viewed on that basis, the present number of people with valid hunting licenses (primary and secondary occupation) correspond to the 1965 figure, where 27.5% of the labour force was employed in hunting and fishing. Based on hunting certificates alone, the number of people authorized to engage in hunting amounts to 40% at certain places (Table 1). In addition to each hunter, members of his family will be employed in the processing of the catch. At the beginning of the century, almost the entire population (13,075 in 1911) was dependent on hunting, a figure which is clearly comparable with the number of people in contemporary Greenland who depend upon subsistence hunting on a continuous basis. Thus, it is safe to conclude that the need for meat, which existed at the beginning of the century, has remained constant.

History

The First Eskimos

Very late in the history of mankind, only 4-5,000 years ago, the vast high-Arctic area of the American continent became inhabitable. It was not until then that hunter-gatherers in eastern Siberia and western Alaska were capable of accessing the rich, but not very accessible, resources — seals, whales, fish, and birds — of the sea ice and the coastal areas of the Polar Basin. To live permanently on the coasts of the Arctic required highly-developed hunting tools, means of transportation, clothing, and housing. Not only did the bearers of the first Eskimo culture adjust to the demands of nature, but during the course of a few generations they spread from Alaska across Arctic Canada to the easternmost area of Greenland.

Archaeologists call these first inhabitants of the Arctic 'Paleo-Eskimos', and unite the various cultures under the designation 'Arctic Small Tool Tradition.' In Greenland, the oldest branch of the Arctic Small Tool Tradition is called Independence I and the Saqqaq culture. Actually, it is only minor nuances in the style of the stone tools which separate these cultures. As indicated by the most recent carbon-14 data, the Paleo-Eskimo hunting communities spread across the Arctic with an astonishing speed. Within a few generations from their point of departure in Alaska, the pioneer groups reached the 'gate to Greenland', the Nares Strait, opposite the Thule district. This happened about 2,500 B.C.

Greenland's rich hunting grounds were also peopled in the course of the same settlement wave. Some settlers found favourable conditions in the northern part, where the muskox was important game. Their settlements were first discovered in Peary Land, where Independence

Janne Jervin

Table 1.
Red and Green Hunting Licenses as a % of Population, 1988/89.

Municipality	Licenses	Population	% of Population
Nanortalik	988	2,653	37.2
Qaqortoq	792	3,436	23.0
Narsaq	518	2,131	24.3
Paamiut	1,036	2,611	39.7
Nuuk	1,872	11,646	16.1
Maniitsoq	1,248	3,992	31.3
Sisimiut	1,175	4,048	23.7
Kangaatsiaq	468	1,263	38.5
Aasiaat	863	3,524	24.5
Qeqertarsuaq	383	1,073	35.7
Qasigiannguit	607	1,778	34.1
Ilulissat	1,221	4,522	27.0
Uummannaq	1,021	2,583	39.5
Upernavik	742	2,229	33.3
Thule	128	745	17.2
Ammassalik	593	2,592	22.9
Illorqortormiut	124	479	25.9

(Average figure for the whole country is 29.1 %).

Fiord gave its name to these people, as well as to subsequent settlers, the people of the Independence II culture. One thousand years later, the latter traveled along the same routes as the first wave of settlers, and used the same hunting grounds.

In between these two groups of settlers, but with the same point-of-departure, and apparently shortly after the Independence I settlers, came yet another group of Paleo-Eskimo hunters who settled along Greenland's west coast, the Saqqaq culture. Shortly after the Independence II group, yet another group of people, the Dorset culture arrived in Greenland. As far as we know, they had the same ethnic origin as the three preceding groups and traveled along the same routes to Greenland. They headed south across Melville Bay and down along the west coast.

Thule and Inussuk

Around the year A.D. 900, the country became inhabited by a vanguard of people who traveled the same routes from Alaska as earlier groups. In the following millennium, this culture, known as Thule, was to take possession of Greenland. We have now reached a period where archaeology begins to overlap with the earliest oral traditions. These people later took the name of Kalaallit, and they are the immediate ancestors of the population of modern Greenland, Kalaallit Nunaat.

The people of the Thule culture were energetic hunters. They either drove away or absorbed the Dorset people, from whom they undoubtedly learned ice-hunting techniques at breathing holes and how to tether harpooned walruses and build igloos.

In Ammassalik, East Greenland, evidence has been found to document that the people of the Thule culture came into contact with the Dorset people of the area. As the Dorset people had probably settled in Greenland about 1,800 years earlier, North East Greenland has been inhabited for a continuous time span of at least 2,500 years.

The rapid progress of the Thule culture was based on the kayak and *umiak* hunting of marine mammals. A comparison between the tools of their ancestors leaves the impression that the latter were giants; their tools were big, fitted with sharpened blades of slate. The most impressive game, which gave the culture its distinctive signature, was the 4 to 60 ton bowhead whale. Whale hunting called for the concentration of many people in large settlements and a social structure whereby a few leaders, who 'called the tune' by virtue of their persona, developed.

The Thule culture in its classic form has only been found in the Thule district. However, the Thule people continued to spread down along the west coast of Greenland, resulting in the elaboration of kayak hunting of seal and small whales concurrently with bowhead whaling. When the Eskimos from Thule settled further to the south and reached open sea, it became paramount for them to be able to hold their own at sea in all kinds of weather. Thus, kayak hunting was developed to perfection, and hunting tools and clothing were adjusted to the conditions of the kayak. Thus, the significant technological features of the culture were changed in such a way that it can no longer be called the Thule culture, but rather the Inussuk culture. Spreading north, people of the

Inussuk culture soon settled Thule, north off Greenland, and down along the east coast.

Almost all hunting was done by means of the kayak, and the dog sled in order to travel far and wide for seals and bears, whereas whale hunting called for different ceremonies. In his book *A Description of Greenland* Hans Egede (1818) describes, among other things, his observations concerning this kind of Eskimo whale hunt. It is justified to conclude that whaling in the following description by Egede had been conducted in the same way during the preceding centuries, or had changed very little:

> When they go whale catching, they put on their best gear or apparel, as if they were going to a wedding feast, fancying that if they did not come cleanly and neatly dressed, the whale, who cannot bear slovenly and dirty habits, would shun them and fly from them. This is the manner of their expedition: about fifty persons, men and women, set out together in one of the large boats, called a kone boat; the women carry along with them their sewing tackles, consisting of needles and thread, to sew and mend their husbands' spring coats, or jackets, if they should be torn or pierced through, and also to mend the boat, in case it should receive any damage; the men go in search of the whale, and when they have found him they strike him with their harpoons, to which are fastened lines or straps two or three fathoms long, made of seal skin, at the end of which they tie a bag of a whole seal skin, filled with air, like a bladder; to the end that the whale, when he finds himself wounded, and runs away with the harpoon, may the sooner be tired, the air bag hindering him from keeping long under water. When he grows tired and loses stength, they attack him again with their spears and lances, till he is killed, and then they put on their spring coats, made of dressed seal skin, all of one piece, with boots, gloves, and caps, sewed and laced so tight together that no water can penetrate them. In this garb they jump into the sea, and begin to slice the fat of him all round the body, even under water; for in these coats they cannot sink, as they are always full of air; so that they can, like the seal, stand upright in the sea... (Egede 1818:102-103).

The encounter with European whalers, beginning at the end of the 17th century, left a profound trace on the Eskimo culture. They introduced gun powder and bullets, iron and wood, bottled spirits, all products easily identified by archaeologists of a later time. But historic descriptions of the Greenlandic subsistence hunt of large whales only show small alterations and variations of hunting methods until the beginning of this century.

Greenland Subsistence Hunting

The 1920's and 1930's were harried by epidemics and diseases, the hunting of various animals failed again and again, and the nutritional and health standard declined alarmingly. In many isolated settlements, people suffered serious malnutrition. During the harshest of these winters, some people starved to death. This precarious situation caused the Danish government to initiate hunting of large whales on behalf of the Greenlanders. This type of whaling lasted from 1924 to 1959, only interrupted by the Second World War.

History of Aboriginal Settlement (continued)

In Northeast Greenland, the settlers continued the traditions of the Thule culture until the 16th century, when new Inuit groups came from the south. The archaeological designation of this culture is the North East Greenlandic Mixed Culture, which, during the 18th century, took on a more local flavour with distinctive houses and tools. The last survivors of this culture were seen in 1823; it is thought that this people was exterminated by the deterioration of the climate.

The Norsemen and the Inughuit

Concurrently, with the immigration of the Thule culture from present-day Canada, Southwest Greenland was settled by a group of Norse farmers, who came from Iceland about A.D. 985. During the Middle Ages, these people made the southwest sections of the country part of the European-Nordic culture area. The Norsemen abandoned Greenland in the 15th century.

The last group of settlers from Canada came to Greenland between 1700 and 1900. They were Polar Eskimos (Inughuit) who have now settled in Avanersuaq (Thule district).

Colonial Era

In 1721, the Danish King sent out an expedition to reestablish contact with the Norsemen of Greenland. Since these people turned out to be extinct, the efforts were directed at establishing a Lutheran mission and a trading post among the Eskimos in West Greenland. Thus, began the colonial era which lasted until 1952. This period did not result in a new wave of settlers, but during the next couple of centuries, Greenland experienced an immigration of Danes, Norwegians, and Faroese. During this time, Denmark established 16 settlement districts in Greenland.

The colonial era, which ended with the outbreak of the Second World War, was a period of isolation that was, to a large extent, a deliberate policy. It goes without saying that Hans Egede, as well as the worldly colonists and traders who followed in his footsteps, were products of their age and they did not have much respect for the necessity of upholding the indigenous culture. But, the colonial regime ran smoothly. At no time was there open conflict or bloodshed. The lifestyle of the Eskimos seemed, to many missionaries and traders, so attractive that they were 'naturalized.' The Danish colonial authorities did not try to control the use of the living resources (marine mammals, birds, fish) which from earlier times had been taken care of by the hunting communities themselves. Aboriginal management consisted essentially of prescriptive rights over specific hunting areas, distinctive to different settlements and families.

The first democratic-political activities of the Danish authority was to place the responsibility for social security in the hands of the population itself, setting up the Boards of Guardians in 1857. Rules governing hunting were not covered by this system — the reason being that this area was already managed by the hunters themselves. In 1908, the process of self-government was strengthened by the establishment of popularly elected National Councils (the so-called Land Councils), one for North and one for South Greenland. The Boards of Guardians were converted into municipal councils modeled after the Danish system. No Danish government officials participated in the meetings of these councils.

As was the case elsewhere in the world, social and political conditions in Greenland changed after the Second World War. In 1950, the two Land Councils were combined into one, the municipal administration was restructured and strengthened, and the scene was set for profound changes.

Integration into Denmark

With the constitutional amendment in Denmark of 1953, colonial status was abolished and Greenland was integrated into the Danish kingdom as a province. Ordinary civil rights were honoured, health standards and education were dramatically improved, and the population grew rapidly. Growing aspirations towards greater autonomy, however, demanded a different arrangement.

The Anthropology of Community-Based Whaling in Greenland

Photos: Klaus Georg Hansen, Nuuk

The whale catchers secure fast to the whale.

The removal of muktuk begins even before the tide has gone out.

The Anthropology of Community-Based Whaling in Greenland

Children look on and learn as a whale is flensed.

Preparing to flense the whale after it has been towed ashore.

The Anthropology of Community-Based Whaling in Greenland

Cutting blocks of meat and muktuk after a successful hunt.

Many local residents participate in the flensing of whales, whether the whale was taken by cutter or during a collective hunt.

The Anthropology of Community-Based Whaling in Greenland

Each participant in a collective whale hunt receives a portion of the catch, according to an elaborate distribution system.

So valued are whale products, that little remains after flensing.

Greater Self-Determination and Home Rule

Act No. 577 of 29 November 1978 introduced local autonomy in Greenland, effective 1 May 1979. This accorded a special status to Greenland as a 'distinct community within the Danish Realm.' Like the autonomous Faroe Islands, Greenland elects two members to the Danish parliament. The Danish Constitution continues to apply in Greenland and sovereignty continues to rest with the authorities of the kingdom. Thus, Home Rule can only be implemented constitutionally by Danish law, with the Danish parliament delegating some of its powers to Greenland Home Rule.

The Act designates to the Home Rule government certain legislative and administrative powers. The basic principle of the Act is that administration of matters of local interest shall rest with the local authorities, whereas matters of more general nature come within the jurisdiction of the central authorities in Denmark. The latter are responsible only for defense and foreign relations, as well as for the overall fiscal policy. Among matters to be transferred from Danish jurisdiction to the Home Rule authorities, only the health service is still pending.

Landsting (Greenland's Parliament) and Landsstyre (Home-Rule Government)

Greenland Home Rule is made up of a legislative assembly elected in Greenland called the Landsting, and an administration led by a small elected government body called the Landsstyre. The Landsting is the supreme political authority within areas delegated to home rule, and, as such, drafts regulations/laws governing areas coming within the jurisdiction of Home Rule.

The Landsting is in session normally 2-3 times a year for a period of 2-5 weeks, once during the autumn and once or twice during spring. The Landsting comprises 27 members, elected directly for a period of four years. At present, four political parties are represented in the Landsting: Siumut (social democrats), Atassut (liberals), Inuit Ataqatigiit (socialists), and Issittup Partiia (liberals). The Home Rule government is elected by Landsting by absolute majority. The premier distributes the responsibilities among the members of the Landsstyre. The Home Rule government is in charge of day-to-day affairs and carries out Landsting resolutions. The responsibility for the administration of Home Rule rests with the Home Rule government. The individual members have their individual areas of responsibility, but all important decisions are taken collectively by the Home Rule government.

Representatives of the Danish State

In pursuance of the Home Rule Act, the High Commissioner is the senior representative of the Danish state in Greenland. The High Commissioner is the supreme administrative authority and in charge of a number of matters of family law, as well as the health service. In addition to the Chief Commissioner's office, certain government institutions are represented in Greenland. The Greenlandic High Court and the Chief of Police, whose jurisdiction encompass the local district judges and the police district of Greenland, are part of the Danish legal system. A naval station under the Ministry of Defense is in charge of coordinating the Danish inspection vessels' activities as well in regard to fisheries inspection as in the matter of upholding sovereignty at sea. In cooperation with the chief of police, it is in charge of search and rescue. In Northeast Greenland, a Danish military dog sled unit patrols the vast expanses of the world's most extended nature reserve. Greenland also hosts a number of American military installations in the defense areas Kangerlussuaq/Søndre Strømfjord and Pituffik/Thule Air Base, as well as a chain of radar installations across the ice cap.

Foreign Relations

The provisions of the Home Rule Act guarantee that the local authorities are to be heard in foreign matters of substantial importance for Greenland. The Ministry of Foreign Affairs in Copenhagen has a special adviser on Greenlandic affairs, and the Home Rule government may assign attachés to Danish embassies abroad. The Home Rule authorities may participate in international negotiations on matters of interest to Greenland, and may be empowered to negotiate on their own without officials from the foreign service.

Greenland is an autonomous customs zone. The *Landsting* stipulates import levies and regulations governing imports and exports. The rest of the kingdom is treated as a foreign trading partner in this respect. In pursuance of the Home Rule Act, treaties and other international agreements requiring the assent of the *Folketing* (the Danish parliament) and particularly affecting the interests of Greenland, shall be referred to the *Landsting*.

Greenland and the EEC
A consequence of the introduction of Home Rule in Greenland in 1979 was that the majority of the local electorate opted for Greenland's withdrawal from the European Community. The reason for this development is clear: following the delegation of powers from Denmark to Greenland, a political struggle was bound to arise over the natural resources of the waters surrounding Greenland. To a population whose survival and economic ability are linked to the marine resources, it was a political problem that these were administered from Europe. The issue was: control over their own resources. Therefore, Greenland opted out of the EEC. Following a referendum in Greenland, this became a reality on 1 February 1985.

Greenland and Nordic Cooperation
The same year Greenland left the European Community, closer cooperation with the Nordic nations was established. In pursuance of the so-called 'Helsinki Agreement', Greenland's Home Rule government now participates in the work of the Nordic Council of Ministers, at the ministerial level as well as in the committees of government officials. The Greenland Landsting has two representatives in the Danish delegation to the Nordic Council.

In recent years, Greenland has cooperated more and more with Nordic countries, in particular in matters of culture and education, as well as in research dealing with multi-species management. As of 1 January 1986, the Nordic Institute for Greenland (NAPA) was established in Nuuk. NAPA's objective is, among other things, to support and promote culture, education, and research in Greenland. Another important task is to strengthen the ties between Greenland and the other Nordic countries.

The Municipalities
Municipal self-government began in 1905. For many years, the most important matter dealt with by the municipal councils was the issue of social security. Expenditures in this connection were reimbursed by the Land Councils and the central authorities.

In the mid-1960's the Land Council placed small amounts of funds at the disposal of local authorities to be used at their discretion for construction works, cleaning, and maintenance of roads, etc. From 1971 a fixed portion of the import levies was allocated to the local authorities.

By 1975, the municipalities set up technical advisory bodies within their administration, and in 1979, formal municipal town planning was organized as well. The municipal councils consist of 3-17 members, depending on the size of the municipality.

Provisions governing the activities of local councils are laid down in the Local Administration Act. Each municipality appoints a finance committee, as well as a social committee, a committee on culture and education, and a labour market committee. Most municipalities also have a technical committee. Almost all small settlements have a popularly-elected settlement council which administers the tasks delegated to it by the municipal council. The settlement council may submit recommendations on conditions in the settlement to the municipal council. The latter, in turn, refers matters affecting the settlement to the settlement council for comments. The settlement councils consist of 3 or 5 members elected for a four year period. Municipal zones and boundaries are illustrated in Figure 1.

Sources of Background Knowledge Concerning the Nature of Subsistence in Greenland

Recent archaeological finds of the Saqqaq culture reveal extraordinarily well-preserved implements of wood, bone, as well as garments. These excavations have been conducted at Qeqertasussuk in Disko Bay. The content of the settlement's rich culture layers reveals a story about how people hunted and lived in Greenland some two and a half millennia ago.

Basis of Livelihood: Animals Hunted

The following table (Table 2) representing 45 species, is a long list compared to lists from other Arctic settlements in Greenland and Canada, and the occurrence of the species varies greatly.

Seals

Faunal analyses show that hunting of ringed seal, and especially harp seal, formed the basis of the settlement's economy (Figure 2). Studies of teeth of excavated harp seals indicate that the seals have primarily been caught during spring migration. The ringed seals were probably caught during the spring season's breathing hole hunt.

Whales

The whales are represented almost entirely by small shaped bone, baleen, and tooth fragments. The baleen comes from bowhead or northern right

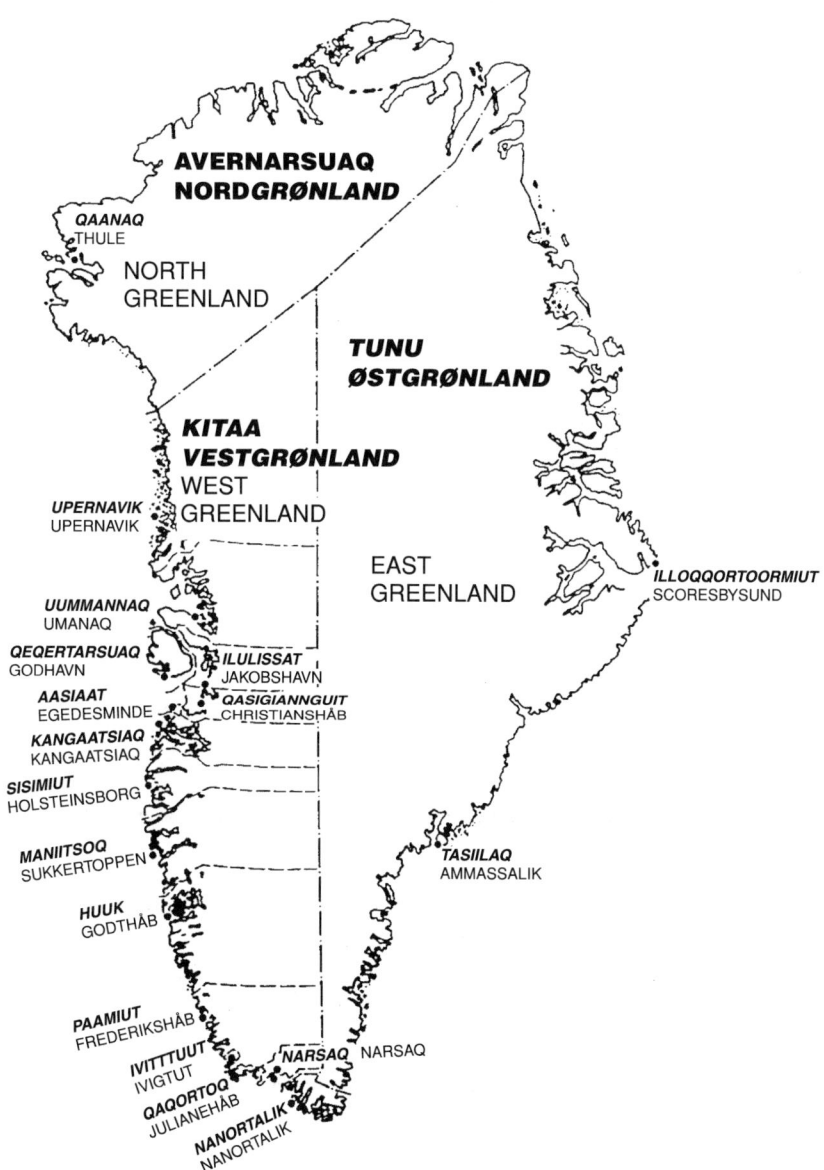

Figure 1. Municipal zones and boundaries in Greenland.

Table 2.
List of Animals Found at Qeqertasussuk.

Mammals	Birds	Fish	Mussels & Snails
Arctic hare	red-throated diver	salmon	common mussel
dog	great northern diver	Arctic char	hiatella bussifera
Arctic fox	fulmar	capelin/ammassat	scallop
caribou	great shearwater	cod	periwinkle
walrus	mallard	polar cod	
harbour seal	eider	sea scorpion	
ringed seal	king eider		
harp seal	red-breasted merganser		
bearded seal	white-fronted goose		
hooded seal	ptarmigan		
sperm whale	Iceland gull		
killer whale	glaucus gull		
porpoise	kittiwake		
narwhal	great auk		
minke whale	little auk		
sei whale	guillemot		
bowhead whale	black guillemot		
	raven		

whales, the bones from the minke or sei whale, tooth fragments from the sperm whale, tusk from narwhal, and various bones from porpoise.

In historical times, Disko Bay was a favourite resort of the largest whales, and it is obvious that it was possible to hunt many whales here 4,000 years ago. There is every indication that whaling was carried out on a large scale, but unfortunately, it is not possible, on the basis of studies of the bones, to conclude whether the large whales were hunted actively, or whether the Saqqaq hunters only cut up floating whale carcasses, which undoubtedly would have been common in the bay. The large amount of tools made of whalebone and baleen (about 70-80 % of the excavated tools), nonetheless, indicate that the large whales played a dominant role in everyday life.

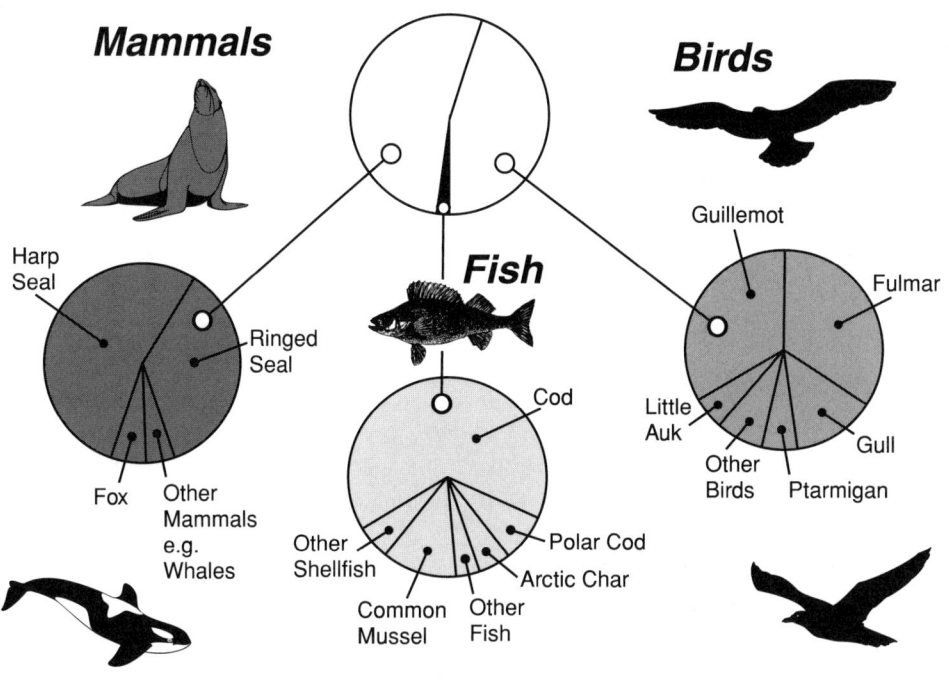

Figure 2. Faunal composition at Qeqertasussuk.

It is not so easy to assess the importance of narwhal and common dolphin. Both species enter the bay during their spring and autumn migration, and there are some good hunting grounds especially north of the settlement. The reason archaeologists find so few remains of these species is probably that only the meat reached Qeqertasussuk. The heavy bones were left at the hunting ground, on the ice or on the beach. However, one thing is certain: the Saqqaq culture boasted active hunters with specialized and handy hunting tools of beautiful workmanship. Some lances were heavy and may have been used for hunting large whales.

Birds

Birds were eagerly hunted. In particular, guillemots, fulmars, and gulls were important. Eider, ptarmigan, little auk, and black guillemot were also frequently part of the diet. The majority of the birds can be caught in the Disko Bay throughout most of the summer season. The conspicuous lack of young birds in the deposits and the occurrence of some species which primarily are found in the autumn migration, however, indicate that many of the birds were caught during the late summer and autumn months.

Fox

Foxes were hunted because of their warm fur, but there are some indications that brain of fox was a delicacy.

Fish

The majority of excavated fish bones comes from large codfish, of which almost all were between 60 and 120 cm long.

Extracts of Reports and Publications Concerning Greenland

The above-mentioned finds, as well as many others, provide numerous examples of the fluctuation of game animals and of the multi-species economy. Later sources confirm that the hunting communities continued to depend on being able to take whatever animals were available. Some of the most copious sources are to be found in the official and annually recurring 'Reports and Publications Concerning Greenland', which provide ample data about the nutritional situation and the public health standard from year to year. 'Reports and Publications' from 1909-1912 describe the period October 1909 to September 1910 as follows:

> During autumn and winter, nutrition was poor in South and North Greenland, and various settlements have received considerable public assistance. The main reason was an unusually poor catch of sea fowl. Especially the eider and in some places the razorbill, failed to appear.

These anonymous sources depict the population's changing health and nutritional situation. The causes of malnutrition are stated to be the weather and the non-occurrence of one or several of the game species, which normally provide food for the autumn or winter months. The period 1924 to 1959 describes whaling initiated by the government on behalf of the Greenland Inuit during a time of nutritional distress.

On a trial basis, the Danish government provided more meat for the indigenous population (1924-1959). As part of this effort, in 1924 the Royal Greenland Trade Department (KGH) commenced whale hunting off Greenland's west coast. Sealing had diminished so much that the people were unable to get the meat needed for themselves as well as for their dogs.

The Sonja

KGH bought the small whaler *Sonja*, built in Leith in 1910, of 127 grt., and equipped with a 300 hp steam engine. It also bought the four-master *Sværdfisken* of 365 grt. with a 190 hp auxiliary engine. The *Sonja* was to catch the whales, and the *Sværdfisken* was to be the mother and flensing ship. During the first four years, whaling was carried out in this way. The blubber was peeled off and salted in barrels which were sent to the oil mill at the premises of KGH in Copenhagen. The meat was distributed among the Greenlanders. People who had participated in the flensing and salting of the whale were allowed to choose first. The remainder of the whale was dragged to the beach were everybody was allowed to take what was left.

In 1928, the *Sværdfisken* withdrew from the project, and the *Sonja* proceeded on her own. After having caught a whale, the *Sonja* would tow it to the nearest town or settlement where it was flensed on the beach at low water. As the meat was meant for human consumption, the period from the time the whale was caught and until it was flensed did not exceed 24 hours. The *Sonja* was a popular vessel. When approaching the chosen port, it was possible at a great distance to tell whether it had caught a whale. If the vessel was listing, people would begin to scream and shout. First the meat was free, but later a symbolic amount was charged. In 1947, the price of the meat was 7 øre per kilogram.

The song 'Sonja Kalippoq' pays a tribute to whale hunting and the amounts of meat it provided. The song about *Sonja* originated in the 1920's and depicts the excitement when a whale is landed in the village. This is the best known and most popular of whaling songs in Greenland.

SUUNIA KALIPPOQ

Pujoq qangattalermat
Kajup nalaatigut
una akisugummat
pujup ataaTigut
/: suaaqattaatileqaat :/
Sunnia kalippoq
immaqa qipoqqaq
/: immaqaluunniit :/
tunnulissuaq!
Pilakkumaartitaasut
niqinnareerpata
utaqqilaartitaasut
atii-inerpata
/: nuannaaginnaqisa :/
Sunnia kalippat
tunnulerujuppat
Niels, Lykars, Makkorsi aamma
pissaqarmata!
neqaa mamannguarmat
tamussinaarlutik
tupittoqartalermat
quniissaluttunik
patiinnallarlugulu
suaaginnartigulu:
Sunnia kalippoq!
tunnulik kamappoq!
/ : tassaqaluunniit : /
asseqassava?
Qujassutissaraaput
kalittoq Sunnia
apeqqutissaraarput
imaattoq tullia:
/ : nukassaannassavinga? : /
qaqugu kalaaleq
Kuukujuk kingulleq
/ : Sunnia nutaami : /
tunnulissava?

SONJA TOWING A WHALE

When the fog was lifting
around the Mount Kaju
the message rang out clearly
across the hills and sea
/: a shout at the top of the voice : /
Sonja's tugging whale
maybe it's a humpback
/ : maybe even : /
a blue whale mighty one!
And when the whale flensers
have got the meat they need
and all the others waiting
have all been called by name
/ : let's all rejoice and be glad : /
Sonja's tugging whale
meat and fat abundant
Niels and Lykars, Makkors also
they have have their share!
Oh, whale meat is tasty
but when they start to chew
a chunk gets down the wrong way
when swallowed clumsily
just pat the guy on his back
just shout at the top of your voice:
Sonja's tugging whale!
the struggle now is over!
/ : was there ever : /
any prey like that!
When Sonja's tugging whale
so thankful we will be
a question comes to mind, though
when turning this around:
/ : for how long shall we be spoiled? : /
when will a Palaaloq
a hero of our liking
/ : on a new Sonja : /
get a blue and mighty whale?

Janne Jervin

In 1933, the whaling captain of the *Sonja* was interviewed about his ventures:

...At first whale hunting was good. We killed about 50 whales a year, providing the population with a minimum of 750,000 kg of meat. It was our policy not to kill more whales than necessary.

...The skeleton was also used — and residual meat was used as dog fodder.

...In 1946 we caught 61 whales, which were distributed among 68 settlements along the coast. It was interesting to watch the flensing. People ran wild; they were anxious to get as much as possible of the meat they hadn't tasted for many years.[2]

The *Sonja's* last captain, Johannes 'Sonja' Larsen, was also interviewed about his adventures onboard the *Sonja*:

... In 1946 we caught about 61 whales during the period 20 June to late October. The whales were distributed among 38 settlements from Augpilagtoq near Cape Farewell and as far as all the settlements in the Umanaq inlet. In the Umanaq Inlet, e.g., I caught a blue whale of 100 — 110,000 kg .... The Umanaq settlement received this whale. There was 50,000 kg of meat and it was 82-83 feet long.

... When hunting whale we could not stay away for too long. A whale has to be cut up within 24 hours, as otherwise the meat is spoiled. If for instance we met a flock of 30 fin whales and if we would have caught two at a distance of 80 nautical miles from the coast, one of them would be ruined after we had towed them to the coast. Thus, it was never a question of ruthless exploitation.

... Whaling was started as a philantropic project in order to secure meat for the indigenous population at a time when sealing had diminished. As meat was a basic element of their diet.... I have often seen that if Greenlanders were given the choice between pork and whale, they always opted for the whale meat....

... I had an interesting encounter when I first started hunting whales in Greenland. It was in Julianehåb and I had gone to bed. A doctor, named Torp-Jensen, paid me a visit and told me that if I wanted to have my name inscribed in the book of life, I should catch a whale for the settlement of Augpilagtok, as there was no place where it was more needed (in actual

---

2. During the war, the *Sonja* was idle.

fact, people were near starving). I told him I had tried in vain, but that I would try again. Having taken in a cargo of 10 tonnes of coal the next morning we sailed at full speed. We passed Sydprøven and approached Cape Egede. North of the Qississut island we met three whales of which we caught one. It was almost dark now, so we called at Nanortalik to take a man onboard as I had never before navigated inside the reefs. We left the next morning as quickly as possible for Augpilagtoq.

Each autumn, the *Sonja* was laid up in Sukkertoppen, and the crew went home to Denmark on the last boat. The following spring, they would return and rig the whaler again. In 1951, the old *Sonja* was sold, and KGH bought a somewhat larger whaler in Norway (250 grt. and 800 hp). It was called *Sonja Kaligtoq*.

All activities were based on the viewpoint that the indigenous population was better off with their natural food, and thus people were warned not to 'lure' them into buying Danish supplies. On the other hand, — as having been the rule ever since 1782, flour and pearled grain could be handed out in case of need, at KGH's account.

In 1936, in his book *Grønland*, Oldendow wrote the following about the government sponsored whale hunting:

> Nowadays the large whales are hunted from KGH's whaler which shoots and tows the whales to the settlements where they are flensed with the assistance of the Greenlanders. As payment they are given the meat and the intestinal fat while KGH keeps the blubber. People shout and cheer whenever the whaler is approaching with its catch; the annual average catch is around 30 whales, mostly fin whales and humpbacks, but also sperm whales and blue whales.

> Economically, whaling is not worthwhile, but it is maintained for the sake of the Greenlanders who get meat for human consumption and dog fodder, bones for fuel and oil for lamps.

Reports on conditions from 1927

The reports state, among other things, that sealing was very unsatisfactory almost everywhere for the six months from April to September:

> At Julianehåb (in South Greenland) April was a rich month as far as hunting was concerned, and prospects were good after the inhabitants had moved to the outer reefs in May.

> Still, in June, the migration of fat hooded seal ceased, and the hunting of lean hooded seal failed completely. Only a few harp seals were caught at

Sukkertoppen, since the migration of seal took place at a later time than usual.

Sealing in the Jacobshavn and Umanak districts has been fairly good, but there were killer whales in great numbers, preventing a good catch, like last year.

The hunt for common dolphin in Upernavik district was poor, and failed completely in the rest of the country.

Due to bad weather, whaling was not initiated until 22 June at Godthåb, and 6 fin whales were caught during the stay there, which lasted until 5 July.

During the period 6 July to 1 August, 4 whales were caught at Sukkertoppen. This poor result was due partly to rough weather — the *Sonja* had to remain idle for 9 days because of a southwestern storm — and partly to the remote location of the whaling grounds as at times the hunting was conducted at Fyllas Banke.

About 1 August the whalers left Sukkertoppen for Godhavn where 6 whales were caught from 4-15 August.

As was the case at Sukkertoppen, the whaling grounds were located far away. At Egedesminde whaling started on 15 August and continued until 16 September, during which period 12 whales were caught — the largest catch from a single ground.

On 16 September the whalers left North Greenland to hunt at Fiskenæsset from 19 to 30 September, blessed with good weather and nearby whaling grounds. Six whales were caught.

From 3 October to the end of the month, 6 whales were caught at Frederikshåb.

Although the weather was good, the catch was not as good as had been expected because the occurrence of whales was low, undoubtedly on account of the large number of killer whales which appeared in the waters with short intervals.

Whaling stopped by the end of October, and the *Sonja* headed for Sukkertoppen to be laid up for the winter. The *Sværdfisken* left Frederikshåb on 1 November for Copenhagen, where it arrived on the 28th. The total catch amounted to: 2 sperm whales, 7 blue whales, 22 fin whales, and 9 humpbacks.

Health conditions, 1927

The population's health condition during the same period was described as follows:

> On the whole, health conditions are not that bad. Nutrition is very good everywhere, mostly due to stocks of dried whale meat, cod, and capelin.

1927-1928:
Reports covering the period 27 October to 28 March tell about an extraordinarily mild winter, also in North Greenland.

> All harbours have been free from winter ice since mid March, and this was unusual. Sealing has been of no significance to the population. Seal hunting in the open water as well as from the ice edge has been poor. Whales were seen everywhere off the coast of Egedesminde during the winter months, and in large numbers during the month of May on the stretch from Kangatsiak to Agto.

> Despite an almost insignificant catch of seal, the nutritional standard was on the whole good, partly due to excellent bird hunting and daily fishing, and partly because many settlements had stocked dried fish and whale meat from the summer's whaling. The standard of health was pretty good during the same period.

1928-1929:

> During the six months from October 1928 to March 1929, sealing was relatively good in North Greenland, but very modest in South Greenland.

> Due to lack of ice, seals could be hunted from kayak in North Greenland (i.e., hunting from breathing holes and ice edge, etc. was not possible that year).

> Nutrition was good, with the exception of the Umanak district. The population here has, since the New Year, found it difficult to make a living due to lack of ice over.

1929-1930:
The Report covers the period 29 September to 30 March.

> Sealing was much better this year in all of North Greenland, whereas it was poor in South Greenland. At Narsalik, in Frederikshåb district, a number of harp seals were caught during the month of January. On the whole, sea bird hunting has been poor in South Greenland where the occurrence of guillemot was unusually low, while eiders were almost non-existent (sic).

But eiders were caught in large numbers at Holsteinsborg as well as in most of North Greenland, compared to the year before.

Nutrition was good. However, South Greenland could not help but feel the lack of sea birds.

Conditions from April 1930 to September 1930

Sealing has been insignificant with the exception of Julianehåb district.

A large number of hooded seal and harp seal were in Julianehåb district. In particular the hunting of harp seal was successful at the most remote islands.

Sealing was insignificant in North Greenland, with the exception of Jakobshavn and Ritenbenk where hunting was not too bad.

Whaling with the 'Sonja' started in mid June at Godhavn and stopped primo November at Sukkertoppen where the ship was laid up for the winter. During the above period, the 'Sonja' hunted whales in the Kisko Bay and South Greenland: total catch: 27 fin whales, 6 humpbacks and 1 blue whale.

Nutrition was good everywhere — and the same applies to the health standard.

September 1930 to March 1931

Sealing was good in North Greenland, but poor in South Greenland. As regards walrus hunting at Agto, the reports say that the catch was excellent this year, altogether 180 walruses were caught, with the Greenlanders succeeding in securing all blubber, hides, and meat from the animals.

Nutrition was good, but a couple of isolated settlements in the Umanak district suffered hardships in January and February because it was impossible to hunt on account of stormy weather and thin ice cover. Communications with the trading posts were also impossible, but the standard of health was good.

April 1931 to September 1931

Sealing was fairly good.

Whaling with the *Sonja* from June to October yielded 16 fin whales, 4 humpbacks, and 1 blue whale.

Nutritional and health standard were good.

Sealing during the winter season was pretty insignificant in South Greenland, with the exception of Qagssimiut and the southern part of Julainehåb

Greenland Subsistence Hunting

district where it was good. In North Greenland sealing was good during the autumn months in Egedesminde, Jakobshavn and Ritenbenk districts, but less good during the last part of the season.

However in the Umanak and Upernavik districts, seal hunters were unsuccessful during the first part of the season.

It has been somewhat better since the New Year. During the summer season sealing was poor in South Greenland, with the exception of Frederikshåb district, where especially Narssalik had a good harvest.

With the exception of Godhavn district, sealing during the summer season was good in North Greenland, in particular at Ritenbenk from June to August with a catch of up to 21 seals per day. Sealing in the northern part of Upernavik district was said to be unusually good.

The *Sonja* went whaling from May to October; total catch amounted to 30 whales, viz. 25 fin whales, 4 humpbacks and one blue whale. Seventeen were caught in South Greenland and 13 in North Greenland.

During the winter season of 1931-32 nutrition in South Greenland was fairly good everywhere. But turbulent weather affected the hunting somewhat. For example, the inhabitants of Sletten municipality experienced difficulties owing to a poor catch of sea bird and seal, and conditions during the winter had been poor in Godthåb.

Nutrition in North Greenland during the autumn months was good everywhere, but the months of December, January and February saw some shortages of food — due to bad weather which has made hunting difficult, if not impossible. Conditions were particularly bad at Uvkusigssat. Qeqertat and Nugaitsiaq in Umanaq district, as well as at Tugssaq, Tasiussak and Kraulshavn in Upernavik district.

The health standard was fairly good during the winter season, but unsatisfactory during the summer season. The east coast (at Angmagssalik) reports almost normal catches of ringed seal. The hunt for hooded seal failed, however, and it is said to be because of lack of polar drift ice.

October 1932 to September 1933

As was the case in recent years, sealing has been of no commercial significance to South Greenland, whereas Sukkertoppen reports a somewhat better harvest than the previous years. Hunting was also good in Holsteinsborg during the part of the winter where the sea was frozen.

In North Greenland sealing has been particularly successful; especially the hunting of ringed seals at Egedesminde district has been better than the preceding 5 to 6 years. The "Sonja" went whaling from May to

October; viz. 21 whales were caught in all: 16 fin whales, 2 humpbacks and 3 blue whales; 11 in South Greenland and 10 in North Greenland.

On the whole, nutrition has been good; in Gothåb, Holsteinsborg and Ritenbenk, particularly good.

Nutrition in Julianehåb district during February and March was unsatisfactory owing to poor cod fishing; at Sukkertoppen it was even worse than the previous years.

Conditions were bad in November at Hunde Ejlande; and they were not so good in Umanaq district during the winter season because turbulent weather interfered with the hunting.

However, *muktuk* from the small whale hunt at Prøven has eased conditions a lot in Umanaq. During the summer season nutrition was good everywhere.

1933 to 1934:

Hunting was good everywhere — at most places better than usual.

The seal and small whale hunt was good all over the country, but in particular in North Greenland. Sea bird hunting and fishing along the coast have resulted in a satisfactory harvest.

Reports from a single settlement in the Umanaq district indicate rather profound hardships at the start of the winter season. The settlement lacks fuel as well as food. The *Sonja* caught 2 blue whales, 2 humpbacks and 24 fin whales distributed as follows:

Godhavn	3 fin whales	1 humpback	0 blue whales
Hunde Ejlande	1	0	0
Egedesminde	5	1	1
Christianshåb	1	0	0
Sukkertoppen	0	0	1
Godthåb	5	0	0
Frederikshåb	2	0	0
Julianehåb	6	0	0
Sydprøven	1	0	0

1934 -1935:

In this season, the whaleboat S/S Sonja brought in 23 fin whales and 6 humpbacks. In wintertime, nutrition has been rather good everywhere, with the exception of the Julianehåb district, where unstable weather and failing hunting results have been prevalent.

Greenland Subsistence Hunting

To the north of Kraulshavn in the Upernavik district, general nutrition has been so poor that the population has had to move south, the reason being that both the narwhal and the seal hunt have failed.

In summertime, though, nutrition has been good everywhere.

In December, Scoresbysund on the east coast has suffered from poor seal hunts. As a temporary measure, people have been allowed to shoot 28 musk oxen.

1935-1936:

During the winter, the health situation has been less than good. The entire population has suffered from a severe and often fatal influenza.

In Ammassalik, nutrition has been good, except for a few months of failing hunt in the Sermilik Fiord as well as during the period of the epidemic.

The health situation in this district has been poor on account of a very severe epidemic of influenza. Sixty three people have died from the disease.

In summertime, nutrition has been good in all the rest of the country.

In wintertime, nutrition has been satisfactory, except in Egedesminde and Godhavn, in Jakobshavn where the ice was constantly broken up by unusual movements of the great glacier, and in Julianethåb where unstable weather and failing hunting results have prevailed. The whaling started by the middle of June in Northern Greenland and continued there until the end of August. By then, the whaler went south along the coast and operated off South Greenland until the middle of October, after which the ship went to Sukkertoppen to be laid up for the winter. The whales were brought in to the following towns and settlements:

Umanak	2 fin whales	2 humpbacks
Godhavn	3	2
Hunde Ejlande	0	1
Egedesminde	1	0
Sukkertoppen	1	1
Frederikshåb	3	0
Julianehåb	4	0
Sydprøven	2	0
Nanortalik	3	0

This report relates the sometimes fatal significance of changing weather conditions for human health and nutrition in the different settlement districts. It details the effects of occasionally failing ice cover which prevented winter sealing and caused near-catastrophic set-backs in the normal food supply. In actual fact, there were several reasons for the reduction of seal stocks in those years. We hear about variable weather conditions which made hunting difficult and dangerous, and of entire species of prey animals which, inexplicably, disappeared altogether from the area for a whole year or more.

In these years we witness the spread of dangerous epidemics like tuberculosis, polio, whooping cough and influenza which took their toll in human lives and left survivors incapable of undertaking any intensive hunting expeditions. In this situation, the hunter-provider assumes a fatalistic attitude: who knows, the traditionally recurring prey of the upcoming season may not show up after all!

Experiences of this character were not new to the Inuit of Greenland. The hardships of former generations had accumulated in local traditions, shaping an insight which enabled people to survive. Nature sets its limits, and you learn to accommodate.

In our day and age, even the smallest settlement has a boat with an outboard motor. And when out, you always bring a gun and ammunition. Even so, you may be up against unforeseeable effects of currents and drift ice, prohibitive weather conditions, and even the total and inexplicable disappearance of the animals you seek. The hunters know that today, as they knew it in the old days.

Well, if everything else fails, these hunters can buy some food for their families in the local store. But, that costs money, and cash comes too as the outcome of a successful hunt. But when that fails..... Money is more needed today than ever before. There is house rent to be paid for, electricity and heating, and gasoline for the boat. (How all this goes together in a viable way of life is dealt with in a following section).

In the decades following World War I, the Greenland population was faced with a string of poor hunting seasons. The climate changed, seals became scarce, other species of prey failed as well, and ice conditions became impossible for normal winter hunting. Malnutrition was rampant, and a decline in the general health condition was becoming a very real threat. At this point, the *Sonja* whaling project was launched. In the Greenland waters, there was no scarcity of good and healthy food. It was only a question of landing some whales.

Aside from the small cetaceans, the traditional small boat whaling only dealt with the so-called slow whales, the humpback and an occasional bowhead. With the *Sonja*, new and abundant meat supplies from healthy whale stocks were brought ashore and malnutrition was countered.

The fact of the matter, of course, is that people in Greenland have always consumed whale meat. In the Arctic climate, the human body craves solid and nourishing food, and many things seem to indicate that imported foodstuff won't do. In the Greenland medical report of 1959, the then chief medical officer, Dr. Preven V. Smith wrote:

> Already the Greenland Commission of 1948 upheld the view that the transition, as an effect of the introduction of the monetary economy, from aboriginal country food to imported foodstuff from the store, is to be deplored. The imported food is simply less nourishing than the local products.

An effect of this report was the creation of the Greenland Council for Correct Nourishment (Ernæringsrådet for Grønland), in collaboration with the relevant nourishment committee in *Godthåb*.

Reports Concerning Greenland (Beretninger vedrørende Grønland), 1955: An Examination of Greenlandic Foodstuff and Eating Habits by the State Health Investigation Board and Vitamin Laboratory

In May 1947, at the initiative of the medical director, and referring to the *Report of June 12, 1946 by the Greenland Committee of the Upper House (Rigsdagen) in Conjunction with a Delegation Appointed by the Greenland Land Council and Representatives from the Greenland Administration*, Dr. Sindbjerg-Hansen from the County of Viborg and Dr. Erik Lynge from the County of Holbæk were sent to West Greenland by the Public Health Board for the purpose of examining hygienic conditions in Greenland. After two years in Greenland, the two doctors produced a report. This report gave rise to deliberations concerning the improvement of health care, which took place in the Greenland Commission of 19 November 1948.

In 1949, a number of samples of common Greenlandic foodstuffs (sea and landfowl, various species of seals and whales, fishes, etc.), were sent to Denmark for scientific analysis. The analyses extended over several years. At one point, another consultant was sent to Greenland for one year for the sole purpose of completing what was considered to be a fully representative array of samples of Greenlandic foodstuff.

The value of these foodstuffs for the consumer can only be fully assessed in the light of the actual composition of the diet in question. It is, however, possible to make an assessment of the nutritional value of each individual element of the average Greenlandic diet by comparing them with internationally-acknowledged tables of the nutritional content of corresponding foods from more southerly latitudes.

It is a well-known fact that Arctic mammals such as seals, whales, and caribou carry thick layers of fat under the skin. Unlike the body fat of non-Arctic animals, blubber and *muktuk* are high in their content of vitamin A. *Muktuk* specifically has been known for centuries as an excellent anti-scorbuticum. But, aside from this, its real nutritional value has not been acknowledged. However, after the analyses, it may now be stated that *muktuk* is a rather rich foodstuff, which, aside from the already known vitamin A and thiamine also contains vitamin C, riboflavin, and niacine.

Meat from Arctic mammals proves to be richer in iron and vitamin A than meat from non-Arctic animals such as beef, pork, and mutton. The other components are equivalent. The entrails from Arctic mammals have a noticeably higher content of vitamin A than their non-Arctic equivalents. The liver of bearded seal may not be eaten because of an excessive (toxic level) content of vitamin C. Otherwise, seal liver is healthy food, and all seal species other than the bearded seal have a vitamin A content corresponding to that of the harp seal, which in this respect is taken to be normative. It should be noted that in all sea species, the content of vitamin A in the liver is higher in older animals than in younger.

The fish in the Arctic contain the same elements as fish in more temperate waters, and their value as nourishment is the same. The guts of both fish and animals generally have a higher nutritional value than the meat.

Summarizing the matter, it may safely be stated that imported foodstuffs contain less nourishing elements per calorie than local Arctic food, especially vitamins A, C, and D, and riboflavin. It is recommended that food habits in Greenland be mainly based on the consumption of local products as opposed to imported foodstuffs.

An Interview with the Nutritional and Dietary Expert, Dr. Peder Helms, M.D., 'Atuisoq', 4 December 1988

In 1936-37 the Norwegian doctor Arne Høygaard examined the food habits in *Ammassalik*, East Greenland. At that time, everybody, at least in the outlying settlements, subsisted exclusively on local food products. All part of the animals were used, even the brain, eyes, and entrails. A chemist, Harald W. Rasmussen, analysed much of the food parts. The daily food habits of one selected person in the settlement of Sermiligaaq were examined. It was a 26-year-old hunter who lived together with his family in a sod hut. He went hunting in his kayak every day, an activity which demanded a lot of energy. Calculations show that his diet contained 16,825 kJ of energy. In order to eat their fill, children have to eat 5-10,000 kJ daily, and adults somewhere between 8,000 and 12,000 kJ. If the adults have to do hard physical work, they may need anywhere from 12,000 to 18,000 kJ to have their fill.

The daily food of this hunter contained 396g of protein, which is a lot. An adult man normally needs only about 70g of protein a day, but one has to bear in mind that in the aboriginal Greenland diet, all energy comes from proteins and fat. When you eat that much protein, the body transforms most of it into carbodydrates, which in turn are being used by the brain, the heart, and the muscles.

The examination of a specific diet in Sermiligaaq in 1936 revealed a content of 120 mg of vitamin C. Out of these, 37 mg came from blubber-pickled blackberries, and 33 mg from *muktuk*, both of which are relatively rare foodstuff. When these are subtracted from the 120 mg, only 50 mg are left. That is less than what experts recommend today, namely 60 mg, and it is a well-known fact that the aboriginal Greenlandic diet mostly was very low on vitamin C. About 25 mg a day is enough to prevent scurvy, and in the old days that amount was secured through the consumption of raw meat and entrails.

There is less calcium in the Greenlandic diet than in most foodstuffs at more southerly latitudes. The hunter in Sermiligaaq consumed 438 mg of calcium, 150 mg of which came from dried capelin. In Denmark, experts recommend 600 mg of calcium a day for adults. However, strictly speaking, 300 mg should suffice, and the Greenland Inuit of the old days are not believed to have suffered any real deficiency in this regard.

As can be seen in the table of 1936, all elements of the diet, except the blackberries, came from the sea. Sea mammal fat is healthier than fat from other mammals. The fats in the Euro-American diet are, to a large

Janne Jervin

extent, responsible for the widespread cardiovascular diseases in the western world. Inuit in Greenland, however, who eat nothing but local food products, never suffer from blood clots or cardio-vascular problems of any kind, although they eat up to 244 g of fat per day on a regular basis.

The reason for this fat tolerance was verified by two Danish doctors, H.O. Bang and J. Dyerberg, during their research among the seal hunters of Illorsuit in Uummannaq municipality. The fact of the matter is that fats from seal, whale meat, and *muktuk* resemble the fats found in fish. Fats from sea animals, whether fish or mammals, turn out to be different from the fats of land mammals. They actually prevent blood clots. The expert advice is therefore to consume meat from seals and whales rather than imported foodstuff. Doctors Bang and Dyerberg conducted their dietary/medical research in Greenland in the 1970's and early 1980's.

The latest dietary scientific investigations support these earlier findings. It is essential that Greenlanders eat as much traditional food as possible. This matter has now been elevated to the highest political level, and dietary recommendations can be expected to form an important part of a general national nutritional policy in the years ahead.

Dr. P. Helms of the Institute of Hygiene, University of Aarhus, wrote about the possible adverse effects of a transition from a traditional diet to a Euroamerican one:

> Any transition away from the traditional Greenlandic diet will entail a risk of introduction of diet-related civilizational diseases in Greenland.

These diseases are the end result of industrially-produced, highly-specialized foodstuffs, with their high content of simple carbohydrates, as well as the high content of fat with a low component of polyunsaturated fatty acids. The recommended total diet should not contain more than 12 g simple carbohydrates per 10,000 kJ. The optimum ration of polyunsaturated fatty acids is 1.0 or above.

Hunting and Subsistence in Greenland in Light of Socioeconomic Relations
(Jens Dahl, Associate Professor, Institute of Eskimology, University of Copenhagen)

The strategy of sea mammal hunting in Greenland is not the strategy of a commercial enterprise nor the strategy of a self-sufficient subsistence economy. This section highlights the complex integrative and sociocultural functions of hunting in the traditional mode of production. The character of socioeconomic relations in Greenland differentiates hunting in that country from commercial whaling and commercial sealing.

Introduction

Every year on the first of May, workers all over the world gather in order, symbolically, to demonstrate and confirm their mutual bond of solidarity. This demonstration is a cultural event with roots in the industrial revolution. The fact that this international event has reached Greenland is indicative of the radical transformation of Greenlandic society since World War II. In Greenland, as in all other countries, these May Day demonstrations are cultural expressions. Home Rule was introduced on May 1st 1979, a date which was originally to be Greenland's National Day. However, this was later abandoned so as not to have the National Day confused with May Day.

Living in an industrialized society, many Greenlanders confirm their mutual bonds of solidarity as modern-day workers. Nevertheless, even families with members employed in the fishing industry rely on traditional subsistence to a greater or lesser degree. Dependent as they are on hunting and fishing, hunters, fishermen, and seasonally-employed workers in the fishing industry often express their respect for nature and for the animals they hunt. These are reflected in cultural expressions, as when the family or the settlement celebrates a young boy's first catch of a ptarmigan or a seal.

Being a worker in the fishing industry, a person celebrates May Day, being a hunter the same person confirms the cultural relations between man and nature. As a worker he cares about wages and working conditions, and as a hunter he protects the distinctive tradition specific to the hunter's mode of exploitation.[3]

The two types of cultural manifestations mentioned above reflect the existence of two modes of production, which, in a Greenlandic context, are inextricably bound together. In this section these modes of production are called commercial fishing, and traditional hunting and fishing.

---

3. Author's note: I use the concept 'mode of exploitation' to stress the contrast with farming societies, where the land is an instrument of labour, and to stress that in hunting societies, man is dependent on nature's own reproduction of the means of life. Thus, 'mode of exploitation' is different from 'mode of production', and in accordance with Meillassoux (1973) and Lee (1981), the two concepts are used as distinct from each other.

Subsistence and subsistence-based hunting, which is particularly concerned with the hunting of sea mammals, form the focus of this section. Taking as my point-of-departure two case examples from the Greenlandic context, I argue that serious problems are generated when our views on subsistence and hunting are dictated and formulated by societies alien to the way of life in question, especially the morals and values of the 'animal rights' movement.

In a global perspective, these societies and organizations are in a politically-dominant position. Therefore, public attitudes toward the role of hunting in Inuit societies develop from the fact that some of these power-holders seem to have succeeded in imposing their values and definitions of subsistence on the Inuit way of life.

Because these views have been advanced by people for whom subsistence has no great importance, Inuit are put in a position with no great influence on what kinds of subsistence hunting are 'acceptable'. Seen from an Inuit point-of-view, the dominant societies only accept certain kinds of hunting, namely, those defined by themselves as subsistence hunting. It is at this point that Inuit (and anthropologists) start searching for acceptable emperically-founded definitions of subsistence. Basically, this approach to defining the role of subsistence elevates Inuit from a dependent position, which leaves no room for an autocentric development strategy based on the integrative role of subsistence activities.

The most widespread notions concerning Inuit hunting traditions are based on a number of fallacies, of which the following seem to be of decisive importance.

Firstly, when discussing Inuit hunting activities, these are often characterized as traditional in ways not related to historical changes, ignoring the fact that today's traditional hunting activities are fundamentally different from those activities labelled 'traditional' 50 or 100 years ago. What should be stressed is that Inuit identity is not attached to any one specific sector of society (e.g., subsistence hunting), and although hunting is labelled a 'traditional' Inuit activity, its historical roots have, in an economic and technical sense, in no way been immutable. Still, it is often regarded as traditional.

Secondly, it is often presupposed that traditional hunting activities and subsistence are incompatible with development, that these activities are traditional because they are outside an imaginary sphere of development and modernization. This is specifically used to the detriment of

Inuit societies when campaigns against seal hunting, fur trapping, and the like are accompanied by non-Inuit demands, which only accept subsistence activities in so far as these are pursued using only 'traditional' means.

Thirdly, the distinction between subsistence hunting and commercial harvesting activities is often arbitrary and fictitious. And, as demonstrated in this article using examples from hunting in Greenland, the quantitative exploitation of wildlife resources takes place irrespective of fluctuations in the world market. However, this does not mean, to take an example, that the collapse of the seal skin market is insignificant in an economic sense. On the contrary, it means that the ecological balance between man and nature remains undisturbed. Furthermore, it should be mentioned that hunting economics in Greenland is predominantly household-related, rather than market-oriented. For many households, money-generating activities are necessary to sustain subsistence hunting, and *vice versa*.

Finally, Inuit culture is often considered to be conservative because it is assumed, erroneously, that what makes Inuit culture different from the dynamic Euro-American societies, is that it is located in a traditional hunting context. Subsistence hunting is an important element in Inuit culture, but the central point is that this culture has developed technical, legal, and political means to cope with the changing economic role of hunting and subsistence. This is illustrated with examples from Greenland, a country where Inuit culture has developed far beyond deep-rooted misconceptions about life in igloos. Before going into the analysis of the harvest of wildlife in modern Greenland, some general remarks would be appropriate.

When Home Rule was established in Greenland in 1979, the new government constituted itself with five members: the premier and the ministers for Trade, Social Services, Education and Culture, as well as for the Settlements and Outlying Districts. The latter post was created in order to intensify the development of those areas which for many years had been neglected, especially with regard to new housing investments.

These areas are those where hunting plays a major or dominant economic role. In fact, the outlying districts are often labelled 'the hunting districts'. Generally, hunting of the non-migratory ringed seal is most widespread, but the hunting of other species of seals, beluga, narwhal, minke whale, fin whale, walrus and polar bear is also economically important. In the central regions of West Greenland, caribou are usually hunted in a period of two months, from August to October. In six of Greenland's 18 municipalities, which account for about 20% of the population, hunting is classified as the most important economic pursuit, although small-scale fishing occupies a steadily-increasing economic position. Outside these six municipalities, hunting occupies a dominant position in many settlements (from 25 to 500 inhabitants).[4]

Following the implementation of the Home Rule Act, legislative power concerning hunting and fishing in territorial waters was delegated to the Home Rule Authorities in Nuuk. This process was furthermore consolidated when Greenland withdrew from the EEC in February 1985. The allocation of quotas and the issuing of licenses and regulations governing hunting and fishing now takes place in Nuuk. The Home Rule Authorities make decisions about various matters, such as the allocation of shrimp quotas in each fishing district, regulations governing the hunting of whales, and hunting seasons for caribou.

In matters of international concern (e.g., negotiations in the IWC), the Home Rule government negotiates through the Danish government since foreign affairs are a matter of national jurisdiction. In affairs of local concern, the regulation of fishing and hunting has, by long-standing tradition, been delegated to the municipal authorities.

As dealt with later, the licenses differentiate between full-time hunters and fishermen and part-time hunters and fishermen, and, finally, sportsmen. Each category has specified rights to hunt, fish and, to a certain extent, to trade the products. This categorizing of people represents an efficient instrument to regulate the harvest of endangered species and of allocating scarce resources between different socioeconomic groups. In this context, it is important to stress that subsistence, including subsistence hunting, has different meanings to different groups of people. Thus, when analyzing hunting and the Inuit communities of Greenland, the focus should not be on subsistence as such, but on cultural, economic, and structural realities.

---

4. The administrative centre of each municipality is called a 'town', the smallest with 500 and the largest with 11,000 inhabitants.

The Structural Importance of Subsistence Activities in Greenland

Subsistence takes a variety of forms, differing from one region to another, from one season to the next, and between households. One factor often emphasized is that products are consumed locally, that all products from seals and whales are used, that no meat is sold, or sold only within the native community. Each of these criteria are intermittently used to define subsistence, and within the national and international political realm of allocating resources between countries and interest groups, emphasis is placed on factors reflecting external pressures with which the natives are faced.

The role of subsistence and, more specifically, the role of hunting of sea mammals is a matter of economic dominance by the European and North American nations, and of cultural dominance that gives these nations the idea that they possess the right to decide on 'acceptable' types of hunting, globally.

As an attempt to redress this position, I have chosen to analyze hunting and subsistence as integrated cultural manifestations. In accordance with this viewpoint, subsistence can be defined as a set of economic activities, and specifically of those aspects necessary to sustain the producers. While the logic of seal hunting in Greenland may seem difficult to grasp for some, taking into consideration the fluctuations in the world market of fur prices and the subsequent decrease in income, it can be explained by the role of subsistence in this type of hunting economy.

Sometimes meat from seals is marketed by the hunter, though very often it is not. But regular or periodical marketing of seal meat is still a subsistence activity. Defined by its structural position, subsistence falls into two categories. In its first, subsistence is an integrated part of the hunting and fishing mode of production, pursued by independent hunters and fishermen who control their own means of production. As an important unit of production, as well as of consumption, the household is perhaps the major economic unit. As a mode of production, household-based hunting and fishing are typically settlement activities. To sustain this mode of production, all but a few households depend on income from paid employment or on social benefits. Although most wage labour in the settlement is on a casual basis, it provides a significant portion of the capital needed for hunting.

In its second instance, subsistence is a subsidiary activity pursued by people who rely mainly on wage labour and by fishermen engaged in commercial fishing. Today, commercial fishing is carried out mainly by fishing vessels and publicly- and privately-owned trawlers. In these sectors, economic dispositions are made outside the household sphere. Subsistence is either casual, as for instance when a seal is caught by a fishing vessel on its way to or from the fishing grounds, or seasonal.

The Traditional Mode of Production

The integration of the various subsistence activities in Greenland relate to their character as seasonally recurrent phenomena. Sealing, for example, is only viable if cash comes in from other sources such as occasional commercial fishing or the selling of meat and *muktuk* from whaling. For the greater part of the year, production and labour relations take place within the subsistence-based economy. Jobs on land are mostly only available on a seasonal basis. Thus, a condition for working on a seasonal basis in the fishing industry is the availability of subsistence and fishing during the remainder of the year. Conversely, subsistence relies on a cash income from various types of activities.

It is meaningless to differentiate between subsistence and commercial hunting as contradictory or mutually exclusive activities. The viability of both rely on integrated structural socioeconomic relations. This interdependent activity structure promotes a flexibility of lifestyle that is responsive to fluctuating hunting and fishing conditions, as well as for salaried jobs, and for hunting products on the world market.

As is known from other societies in the Third World, subsistence agriculture is necessary for the reproduction of labour, and this also holds true for the traditional mode of hunting in Greenland. Adding to this the fact that cash income is necessary to buy and operate the technological means of production (boat, outboard engine, petrol), required in carrying out subsistence hunting, the intertwined organization of subsistence and cash-generating activities should be sufficiently clarified.

The subsistence based mode of hunting has yet another characteristic which demonstrates its flexible and accommodating structure. In contrast to agricultural societies, which are dominated by exclusive control of farming areas, the Greenlandic hunters and fishermen can turn from paid employment to fishing or from fishing to subsistence hunting from one day to the next, or from one season to another. At least this is the tradition, although an increasing number of restrictions have been introduced in recent years. Since this tradition is of fundamental significance to most Inuit societies, some remarks on the regulations of and access to the pursuit of hunting and fishing within the traditional mode of production should be made.

Hunting grounds and game animals are open to utilization by everybody on a communal basis, subject to the municipal allotment of hunting licenses. This right to the land and sea is linked to use and implies no fee simple title or ownership right. In fact, time has shown a *de facto* expansion of the non-exclusive collective right to hunt and fish, and today no community enforces, nor is in a position to enforce, exclusive usefruct rights to a defined area *vis-à-vis* inhabitants from other communities.

In recent years, industrial encroachment has compelled both Greenlanders and other Inuit to enforce communal and collective traditional hunting rights. This was done, for example, in the spring of 1975, when hunters from the settlements in the Uummannaq municipality physically blocked an icebreaker en route to the lead and zinc mine at Marmorilik. They demanded respect for traditional and non-written aboriginal hunting rights. Their action achieved recognition to such a degree that the future of ice-breaking was to be regulated by agreements negotiated between the mining company and the local population (see also Dahl 1977, 1985b:xxx).

Side-by-side with this flexibility and communality in ecological relations, we find individual control and ownership of the primary means of production. In an historical perspective, this represents one of the most conspicuous continuities in the development of the traditional Greenlandic hunting society. Each individual hunter controls the primary process of production (hunting, fishing) and also the means of production, among which the most important are boats with outboard engines and, in a few areas (primarily in Avanersuaq/Thule) kayaks too. Thus, an historical change in the technical development of hunting equipment can be noticed, but also a remarkable continuity in the individual hunter's control of the hunt as such, and of the means of production.

Since 1979, Home Rule authorities have supported the traditional mode of hunting and fishing through investments in small, decentralized fish-processing units and cold stores. These serve the local trading of different varieties of fish, *muktuk*, and whale and seal meat, later to be sold all over Greenland. These units of production are owned either by the Home Rule agency Greenland Trade (KNI) or by settlement-based co-operatives. Today's decentralized commercial fishing, in combination with hunting of sea mammals, facilitates decentralized habitation and prevents over-exploitation of sea mammals.

As has been mentioned above, rising prices for hunting products or decreasing numbers of animals caught will not necessarily entail intensified hunting efforts. As far as seal hunting is concerned, this is primarily explained by the structural position of this activity in Greenlandic society and economy. To this should be added, as a general statement, that when it comes to the interaction between the different social strata, hunting regulations are clearly facilitated by the combined effect of dealing with an open ecological system, a decentralized pattern of habitation, and a very limited geographical range of activity.

As a way of life, hunting is an activity concerned with the present moment, and there is no reason to believe that hunters in a modern setting will always try to protect renewable resources with a longer time perspective in mind. Over time, it actually has happened in Greenland that local populations of non-migratory species (Arctic char, caribou) occurring in closed ecological settings, have been eradicated.

To counteract such developments, the Greenlandic authorities (the municipalities and Home Rule government) have, for many years, regulated those kinds of hunting and fishing which could endanger local populations of game and fish.

In general terms, and concluding the issue, it should be stated that wildlife protection measures taken by the autonomous and/or municipal government, have shown themselves at least as useful as international regulatory measures. Besides, as a motivating force, they are more efficient than the latter, stemming as they do from authorities closer to the hunters themselves and more intelligible to them.

Hunting and Fishing and Subsidiary Activities

Many people with low salaries or with only seasonal or part-time employment rely on hunting and fishing as important economic and social activities. One is justified in stating that in this context, the hunting of sea mammals is altogether a non-commercial venture. Being only a subsidiary activity, hunting in this case nevertheless assumes an important role in the reproduction of labour, due to low salaries and periodically recurring seasons that offer no employment. Owing to its reproductive functions, this type of hunting should also be labeled 'subsistence'. Its commercial significance is negligible, as described under the traditional mode of hunting and fishing, and its structural role is different. This is no 'dual economy', but an arrangement necessitated by this specific economic situation (Dybbroe and Møller 1978: 14).

Contrary to what is widespread in many agricultural societies, where the men monopolize the salaried jobs and the women are left to take care of all subsistence activities, in Greenland the distinction between wage labour and subsistence activities is not identical with the sexual division of labour. Most subsistence activities in Greenland are performed by men. As a consequence, a woman cannot take over in the event of her husband turning to some kind of wage-earning activity. Only in extreme cases will a wife be observed to take one of her grown-up sons with her in the boat, just to jig a few fish for dinner. In those cases where the sexual division of labour corresponds to distinct economic activities, most often it is the women within the household who has the paid employment, and not the man.

However, hunting and fishing are increasingly reserved for specific groups of people and the purpose is not so much to protect or favour subsistence activities, but rather to comply with demands made by so-called 'full-time hunters and fishermen'. As it is, all people engaged in hunting and fishing today belong to one of the following three categories.

Full-time hunters and fishermen are those people whose primary source of income (in cash) derives from hunting and fishing. This group includes independent (self-employed) hunters/fishermen, and persons employed on fishing vessels and trawlers. Only these persons are allowed to hunt, for instance, whales and, in some areas, caribou. In other districts, they have preferential right to harvest, for example, caribou. The trend is that the procurement of an increasing number of species of fish and mammals is reserved for people belonging to this category. Through the Greenlandic Hunters' and Fishermens' Organization (KNAPK) they try to influence the political process. In periods when too many fish are landed (beyond the capacity of the fish-processing industry), people registered as full-time hunters and fishermen will be given preferential status to trade their catch over those for whom hunting and fishing is a secondary activity.

People who have their main income from paid employment (on a seasonal or year-round basis) are designated 'part-time hunters and fishermen'. Many of these people receive their main income from seasonal employment in the fishing and construction industries, or other kinds of wage-earning jobs. They also have the most pressing need for both non-cash income (subsistence) and cash income (selling of Greenland halibut to Greenland Trade, for example in lean periods). It is interesting to note that, as well as being members of a trade union, many of these people are members of the Greenlandic Hunters' and Fishermens' Organization.

The third category consists of sports fishermen and hunters. Danes, primarily, belong to this category.

In conclusion, it is worthwhile to emphasize that, in the management of hunting activities, a transformation has taken place from a condition where man individually exploited nature around him to a state where the sometimes conflicting interests of different social and economic groups are combined and organized. Today, the regulation of hunting and fishing is under Home Rule jurisdiction.

One thing remains certain: the distinction between subsistence and commercial harvests is artificial. The two phenomena intertwine to form a distinct mode of production. Maintaining a distinction between the two only serves to undermine local control, while strengthening neo-colonial control, especially in regard to establishing the status of and protecting endangered species. The question is not whether endangered species should be protected. Everyone agrees that they should. The question is how it should be done.

Conclusion

Sea mammal hunting in Greenland is not the strategy of a commercial enterprise, nor one of a totally self-sufficient economy. Likewise, we know that the strategy of cattle raising among the pastoral Massai of East Africa is quite different from that of a capitalist cattle ranch. The difference is the variable cultural significance of cattle raising as well as of hunting. I have tried to highlight the complex interactive and cultural functions of hunting and fishing in the traditional mode of production, and how they articulate with the capitalist mode. The character of these socioeconomic relations differentiates Greenlandic hunting from commercial whaling — the Europeans nearly exterminated the bowhead whale from the waters of Greenland — and from commercial sealing as it used to take place in Newfoundland.

In separating the traditional mode of hunting and fishing from industrialized fishing, stress has been put on the role of hunting in reproducing relations of production. Defining subsistence as activities of reproduction, two very different pictures have emerged. One challenges hunting as a recurrent and continuous activity. The other sees hunting and fishing as subsidiary activities by persons employed in salaried positions. It has been shown that the oft discussed distinction between commercial vs. non-commercial is meaningless and artificial in a Greenlandic context.

Accepting self-determination by Greenlanders can only mean that responsibility for resource management ought to stay where it is, i.e., with the settlement, the municipal council, and the Home Rule government.

Traditional and Present Distribution Channels in Subsistence Hunting in Greenland
(Robert Petersen, Professor, Greenland University, Nuuk)

The Homogeneous Community

The traditional community in Greenland consisted of independent and self-supporting households. Households were autonomous, economically as well as organizationally; the various demands of the households were met by the household itself, and the different chores were done by the household members. In this way, individual households were placed on an equal footing. All of them produced the same things, and thus the need for barter between households in the same settlement or area was rather limited. Strictly speaking, the sharing of prey was not an economic necessity, but a kind of economic and social insurance (Petersen 1987: 11). Every household gave shares, when possible, to other households, and received shares in return.

Generalized exchange had exceptions, even in traditional Greenland. This kind of generalized exchange (cf. Sahlins 1965: 147) was common between different households in the same area; it occurred between groups with a realistic possibility of daily exchanges of like products. There are, however, two groups who, even in those days, could not join generalized exchange relationships: individuals outside the settlement households, and households from different areas (Peterson 1987:12).

Even in homogeneous communities, you would find individuals who did not enjoy the same opportunities as everyone else, e.g., individuals outside the households proper, incapable of meeting their own needs by their own activities. These people were not able to give things corresponding to what they received. They shared the distribution of food within their community, but had to pay for other necessities, either by giving other kinds of products in return or by offering some kind of work (*ibid.* 12).

Another kind of exchange occurred between households from different areas. Here, barter was an important part of the exchange between people belonging to different groups. An essential factor in this kind of exchange was the different ways of distributing the resources, combined with the lack of possibility of reciprocation with daily exchanges.

Solidarity and Balanced Exchange

It is often said about exchange in the traditional Greenlandic community that people gave without thought of 'payment'. This is not completely true. But it is true that, normally, they did not mention payment. There was a generalized exchange of prepared food when people were invited for a common meal (Rink 174: 33), or when people invited you for another common meal. Nobody spoke of 'payment' as this kind of sharing was part of the common behavioural pattern in the community. There was a natural expectation of receiving something equal in return. In fact, you did not give to people things other than food to be consumed, without any possibility of them giving the same things in return. But although this was standard procedure, people normally did not demand anything in return.

A kind of group solidarity may be found in the exchange of goods. For example, if you helped a relative, it was possible that your aid could be repaid when the same person helped another relative of yours (Petersen 1987: 28). This sort of solidarity is often connected with non-material help.

Balanced exchange

If, for instance, you needed a seal skin, but were unable to catch seals yourself, you might acquire the skin by buying it. You could repay by giving fish in return, or you could help people with covering their boat, or you might fetch something for them. Even story-telling could be 'exchanged' for food (Lynge 1957, I: 7). People also considered magic songs as a kind of property, but as magic songs were secret, there was probably no direct exchange in this domain; if they were publicly known, the songs might lose their power. In case you needed a magic song, it could be because you had none yourself, so that you had to pay for it with something else (Rink 1974:51). Also, the services of an *angakkoq* (shaman) usually required payment (Rasmussen 1924:224).

The underlying concept of this system was that, between households from the same settlement, there was a common exchange of food, in which every household was capable of giving as well as receiving. There was no hierarchy of households; all were, by definition, equal to each other. In case you were not a member of any household, and you needed some kind of goods, even some local products, then you had to pay during the transaction, according to the value of the thing you bought.

Sharing game
In connection with sharing game, exchange is still more generalized. This is probably owing to special norms. When it comes to the large mammals, such as narwhal, beluga, walrus, and polar bear, a person might well be recognized to be the one who actually caught the animal, but the other households had a right to shares of the prey just by being present before the prey was brought home.

Sharing of game as a distribution channel
The situation whereby individuals lived outside the households proper also demanded a kind of exchange. The barter between households from different areas indicate the same thing, and paying for magic songs, story-telling, or shamanistic help also reveals that, even in traditional Greenland, remuneration by people who could not give the same type of things in return as they received was acceptable.

When describing the sharing system connected with large game, it is vital to stress that it also had another function. It worked as a distribution channel. A positive side-effect of the system was the idea of wise utilization of renewable resources: it was not necessary for all the households in a small settlement to try and catch their own animals, when one, or a few people, might meet the demands (Petersen 1983).

Development of the Greenlandic Community in the 20th Century

At the beginning of this century about 11,000 persons lived in Greenland. The population mainly consisted of hunting and fishing families, with the main emphasis on hunting (Amdrup *et al.* 1921). But even then there were wage-earning families engaged by the state, either at the Royal Greenland Trade Department or with the school and/or health services. The important thing in this pattern is that the households receiving shares had the same possibility of catching the same kinds of animals some time in the future, and thus of giving like shares in return. It was of no real importance whether they would actually succeed in doing so.

Around 1960, a West Greenlandic community reformulated the rules for sharing of game. Steps were taken to exclude fishermen from the exchange system; the fishermen could only receive shares, but never contribute to the system (Kleivan 1964:69).

Common exchange had important social effects. By exchanging goods and services you supported solidarity, and might in this way contribute to a friendly and peaceful atmosphere in the local community. In communities without any organized public insurance system, this kind of sharing was a very important factor. During times of distress, as well as in old age, exchange and sharing constituted an acceptable form of insurance.

In the traditional Greenlandic hunting community, a pool system was unknown. Inuit hunting was characterized by an individual's acquisition of goods. Even in organized hunts, rights of access were individualized. Therefore, property marks of different kinds existed in the Inuit communities where organized hunting was developed (Boas 1899). When archaeologists find hunting implements with property marks, this means that organized hunting occurred (Meldgaard 1986:31).

Today, the Greenlandic population of Inuit origin numbers more than 40,000 (Anon. 1986:24). Of these, about 11,000 persons live on fishing and hunting (cf. *ibid.* 30). The rest of the adult indigenous population are wage-earners of different kinds, and to various degrees. Thus, about 3/4 of the Greenlandic population today has no possibility whatsoever of joining the general exchange of hunting products. In this way, they correspond to those individuals in the traditional community, who lived outside the household structure, or to the households living widely apart, without possibility of daily exchanges.

Exchange and Distribution of Goods Today

Theoretically, if general exchange was the main principle for today's distribution, communities might experience peculiar and undesirable situations. For example, the general exchange might stop altogether, so that people within hunting and fishing households would end up with great quantities of different kinds of meat, while the rest of the Greenlandic people, about 3/4 of the Inuit population, suddenly had to live from imported food alone. Alternatively, the hunting and fishing groups could continue to provide meat to the other households, expecting, unrealistically, also to receive meat in return. This, too, would bring about a situation that would be untenable locally (cf. Hertz 1977:16).

To forestall this dilemma, a system has developed whereby everyone contributes to distribution channels and exchange relationships within their own means, some with their own products, others with services, and still others with money. This is why, in the end, money had to enter the exchange system if it was to be kept open at all.

The Role of Money and the Commercialization of Hunting

In the international and false discussion about the difference between subsistence and commercial hunting, some have maintained that no definition of hunting as subsistence is possible if money is put into the distribution channels (Petersen 1983). In the international debate, however, the rejection of commercial hunting was not just based on money entering the system, but on the size of the catch being determined by a profit maximization. In commercial hunting proper, investments not only call for more efficient hunting methods, they also require new investments and create a need for still more income.

Nothing of the sort is seen in aboriginal subsistence hunting, even if distribution of the game secured requires the exchange of money. There is no profit maximizing mechanism and no ensuing growth in the pressure exerted on the resource in question. The role of money in the distribution channels does not justify the labeling of aboriginal subsistence hunting as commercial in the sense used by the IWC when dealing with 'commercial whaling.'

Implements Used for Hunting

In this context, it may be relevant to evaluate the implements used in a hunting expedition according to their effect on the result. These include implements used for:

- transportation from home to the hunting area and back (a)
- the hunt itself (b)
- transportation of the prey (c), and
- butchering and distribution of the prey (d)

Whether transportation is by boat or sled, improvements of this kind have had no direct influence on the hunt itself. The same is true for (c) and (d). Similarly, a tractor for pulling up the prey before flensing, or other mechanical improvements has no effect on the species hunting. Of course, by saving time in the long run, these improvements cannot avoid making the enterprise of hunting more efficient. Improvements in the hunting implements proper (b) could have an immediate influence on the predation rate of animal populations. However, the best kinds of improvements would not necessarily be those that would make it easier to obtain more animals, but those which make it easier to secure the prey. The principle of 'least effort' and shortages of labour to process and distribute the catch, not to mention a deep respect for hunted animals, combine to limit the amount of animals taken at any one time.

This touches on the question of money, which is also needed in connection with subsistence hunting. In traditional whaling, relatively great investments in technology, skill and labour, were necessary (Gamble 1984:39). As in earlier times, it makes little difference whether the crew was paid in food or in money. So, in reality, it isn't a question of whether or not money changed the motivation for hunting. The important issue is whether or not the money channeled into hunting brings about an increase in prey secured and capitalistic motivations.

Some Aspects of Distribution

In any Greenlandic town today, you will find an open air market for fresh products. Here, hunters and fishermen sell their catch, normally caught the same day. You might buy fowl and seal meat, different kinds of fish, and meat from small whales. You may also find caribou meat and edible plants according to season. Wage-earning households buy the fresh local food they need, and hunters and fishermen sell their products and exchange their hunting and fishing experiences.

Products from this market may also be sold to local retailers. In this process, refrigerator technology plays an important role. Veterinary and sanitary considerations are stressed by health services and consumer groups. The leftover products are bought by local processing firms, co-op chains or publicly-owned stores. In this way, one is able to market hunting products in the hunting area proper, as well as in the towns of central West Greenland.

Exports

Greenland exports are based upon fish products, minerals, and seal skins. No whale products are exported. An exemption of a kind is made for indigenous Greenlanders living outside their home country. They are occasionally provided with whale meat on a non-commercial basis.

Subsistence or Commercial Hunting?

Different definitions of aboriginal or subsistence whaling have been formulated in connection with the work of the IWC. In 1931, a definition was formulated in which the use of canoes, pirogues, or the like; the use of firearms; the employment of non-native persons; and the delivery of whaling products to any third person, was prohibited. Today, definitions of this kind are regarded as oppressive.

Later definitions containing the ideas of local consumption by the aboriginals were somewhat unclear because the term 'Aborigines' was not defined. The idea changed from only accepting hand-held harpoons to increasing prohibition of non-explosive projectiles (Government of Japan 1987). One of the definitions used to describe 'subsistence hunting' is connected with the concept of 'local consumption'. 'Local consumption' is defined by Finn O. Kapel in three paragraphs:

- Consumption of prey animals by participants of the hunt in question, and by the families of the participants.
- Consumption of the prey animals by other households within the same town/settlement.
- Consumption of the prey animals by people living in other places, but with familial and cultural relationship to the hunting groups.

All of these three elements are necessary, when we define subsistence hunting in relation to the concept of 'local consumption.' Thus, according to Kapel's definition, hunting is not commercialized as long as the prey is consumed by the families of the hunting groups, by their settlement mates, or by others with a familial or cultural relationship to the hunting group.

This definition is probably valid in connection with the small population that characterizes the hunting communities. Besides, Greenlandic hunting, even today, is still a low-technology and low-income occupation, where initial capital investment rates are low. Only whaling calls for high-technology and only in as much as the penthrite grenade is coming to the fore.

Symbiotic Relationship

Despite its name, Greenland is not an agricultural nation. Living in this country is based upon the use of marine and wildlife resources. Hunting households are the primary consumers of their own products. In the dog sled region, both human beings and their dogs utilize the hunting products. In fact, hunting products are rather expensive, and the hunters can afford to utilize their own products (Petersen 1983) only because they need not pay for them. For the hunting communities, the consumption of their own products is essential, as the hunting products are better than agricultural products for making people fit for the cold climate. Also, wage-earners may continue to hunt on the side. In this way, many facilities have a welcomed extra source of income. But hunting is no obvious alternative to 'landlubber' jobs; it demands specific investments and may prove an expensive and dangerous way of life to people who are unfamiliar with the land and the sea.

Thus, we find two main groups in Greenland living side by side in a symbiotic relationship. The smaller group of hunters and fishermen who produce more food than they are capable of consuming, and a large group of people who need and appreciate hunting products, but who are unable to hunt and fish for themselves. The latter group only has one reasonable means of exchange, money. By using money, they acquire the hunting and fishing products they need, and the hunters and fishermen can in this way continue a reciprocity with other households.

Free Distribution Today

Occasionally, free distribution occurs even today. Especially meat gifts occur, not least between relatives, and between people with possibility of maintaining exchange parity. One may also receive meat-gifts as an acknowledgment of help received. Personally, I have had the opportunity to see people bring meat gifts for the family of the clergyman. In this way, the clergyman's work on your soul was an object worthy of consideration.

Basically, the traditional channels of distribution are kept open in many ways. But society is no longer a homogeneous community and, as mentioned above, even in the so-called traditional homogeneous community there were population segments between whom a free exchange was limited, and where the barter-like situation replaced free and generalized exchange.

In contemporary Greenland, a free exchange of hunting products is impossible as a general and comprehensive rule. The rules governing the sharing of prey have had to be regulated according to a consideration of a reasonable balance in the nature of exchange. Under present-day circumstances, this entails the introduction of money, which is the only means that can prevent the distribution channels from collapsing.

Conclusion

Even in the traditional Greenlandic hunting community, a generalized, if indirect exchange of products occurred only between people with equal possibility of giving and receiving the same kinds of products. Between people who had no possibility of exchange parity, a barter-like exchange occurred, especially when somebody wanted goods which were not included in the daily food distribution. This kind of balanced exchange could be an exchange of hunting products for other kinds of products, or hunting products for a service rendered, or a service of a like service (e.g., in connection with covering of an *umiak* with new skins, as this kind of work demanded cooperation).

Special kinds of help (e.g., aid from a specialist of magic) also demanded balanced payment, because the partners had no possibilities of both giving and receiving the same service. Today, meat gifts for a clergyman or a teacher might be said to serve the same purpose. Meat gifts, and especially traditional rules for the sharing of prey, may be regarded as a kind of an insurance system that would make it reasonable to receive different goods during difficult times. In addition, sharing rules play an important role in connection with the idea of wise utilization of natural resources. An important factor is that the system of sharing rules is valid only between people with equal opportunity to reciprocate.

Heterogeneous communities made it necessary to accommodate the idea that people incapable of contributing to the system in a traditional manner had to find other means to participate. This is an important consideration, as hunters often belong to the low-income group. But, it is also a consideration implying that the taking and giving system must be somehow reasonable and feasible for everyone.

Today, where 3/4 of the indigenous population still appreciates the products of the hunt, but have little opportunity to hunt, it has been necessary to introduce money into the exchange system. Money has permitted the channels of distribution to remain open, without abandoning the traditional system in which people with different opportunities are expected to contribute to the system by their own means. This has no real effect on the animals hunted and the money is not used for maximation of the hunting products. Its primary function is to keep hunting alive and to provide people with good, inexpensive, nutritional food. Neither Greenlanders, nor anyone else, should regard this development as anything approaching the commercialization of hunting in Greenland.

References

Amdrup, C.C. *et al.* 1921. Grønland i Tohundredeåret for Hans Egedes Landing I. *Meddelelser om Grønland*,vol. 60.
Anonymous. 1986. *Grønland 1986*. Ministeriet for Grønland.
Boas, F. 1899. Property marks of Alaskan Eskimo. *American Anthropologist*.
Dahl, J. 1977. *Minedrift i et fangersamfund*. Verbaek.
Dahl, J. 1988. Mining and local communities. *'Etudes/Inuit/Studies* 8(2): 145-157.
Dybbroe, S. and P. Møller. 1978. Fangstens betydning i dagens Grønland. *Folk* 13(1): 7-15.
Gambell, R. (editor). 1984. Aboriginal subsistence whaling. *International Whaling Commission, Special Issue 4,* Cambridge.
Government of Japan. 1987. *History of the consideration of aboriginal subsistence whaling,* Tokyo.
Hertz, O. 1977. *Okologisk undersøgelse af minedriftens virkning pa fangerne ved Uvkusigssat*. Kragestedet.
Kleivan, H. 1964. Acculturation, ecology and human choice: Case studies from Labrador and South Greenland. *Folk* 6(1), Copenhagen.
Lee, R.B. 1981. Is there a foraging mode of production? *Canadian Journal of Anthropology* 2(1): 13-19.
Lynge, K. 1957. *Kalatdlit oqalugtuait oqalualaivilo*, I-III, Nuuk.
Meillassoux, C. 1973. On the mode of production of the hunting band. In *French perspectives in African studies*. P. Alexander (editor), Oxford.
Petersen, R. 1983. The Greenland Inuit culture, Greenland Home Rule: Why does Greenland want to leave the EEC? Nuuk.
Petersen, R. 1987. *Nunatta oqaluttuassartaanit*. Nuuk.
Rasmussen, K. 1924. *Myter og Sagn fra Grønland, I-III*. (1921-1925), København.
Rink, H.J. 1974. *Tales and traditions of the Eskimo*. London and Copenhagen.
Salhins, M.D. 1965. *On the sociology of primitive exchange, the relevance of models for social anthropology*. ASA I, London.

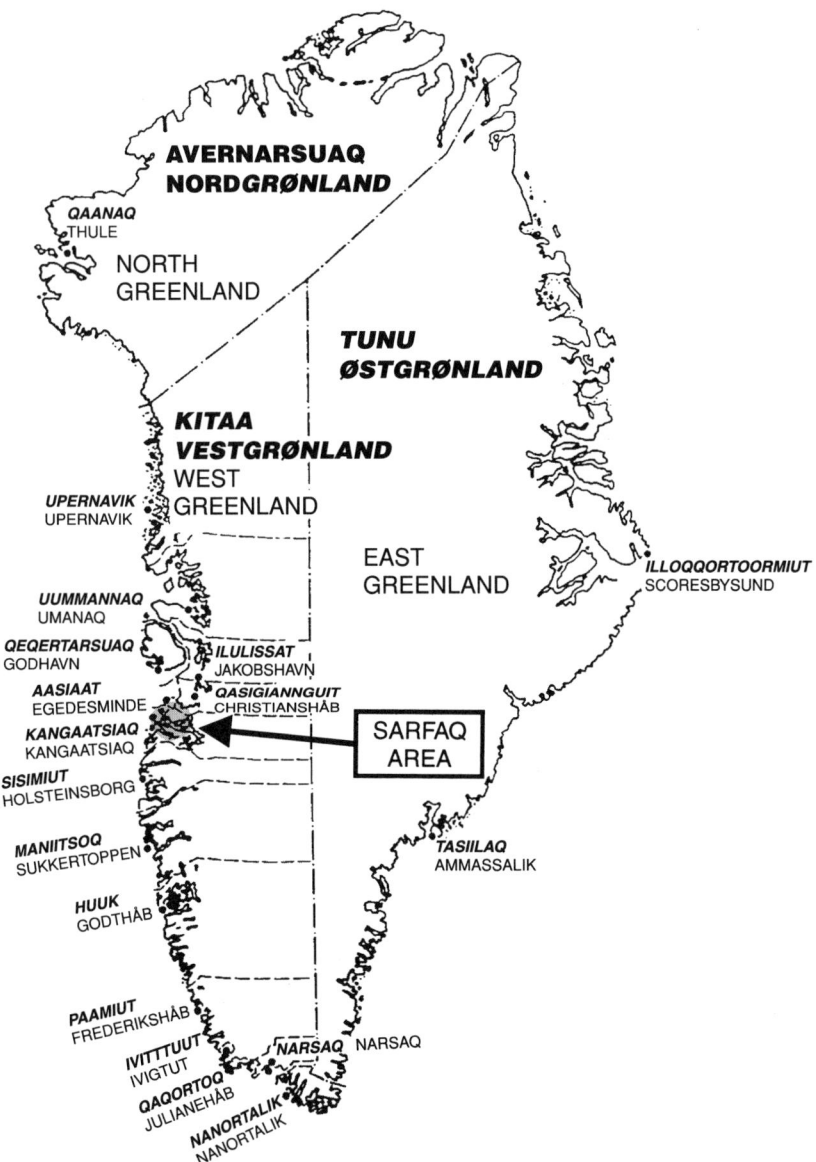

Figure 1. Sarfaq area on west coast of Greenland.

Inuit And Whales at Sarfaq (Greenland)

Svend E. Larsen and
Klaus G. Hansen

prepared for the
Greenland Home Rule Government
at the occasion of the 42nd Annual Meeting of the
International Whaling Commission
1990

(Ode to the whaleboat *Sonja,* by Peter Olsen, 1892-1930)

Pujoq qangattalermat
Kajup nalaatigut
una akisugummat
pujup ataatigut
suaaqattaatileqaat,
suaaqattaatileqaat:
'Sunnia kalippoq.
Immaqa qipoqqaq,
immaqaluunniit
tunnulissuaq.'

Pilakkumaartitaasut
neqinnareerpata
utaqqilaartitaasut
atii inerpata
nuuannaaginnaqisa
Sunnia kalippat
tunnulerujuppat
'Niels, Lykars, Makkorsi' aamma
pissaqarmata.

Svend E. Larsen and Klaus G. Hansen

Sarfaq

Niaqornaarsuq is the second largest of four settlements that make up the inhabited part of Kangaatsiaq municipality. The administrative centre, the village of Kangaatsiaq, after which the municipality has been named, is the largest. The municipality covers an extensive area of skerries and fjords in central west Greenland (Figure 1). It stretches for some 130 kilometres from the Nassuttooq Fiord, which forms its border with the neighbouring municipality Sisimut to the south, up to the northern part of the Clay Plain, Naternaq, bordering on the neighbouring municipality to the north, Aasiaat. In the other direction it reaches from the inland ice in the east to the archipelago in the west, a distance of some 175 km.

The Niaqornaarsuq settlement is situated on a peninsula in one of the two large systems of fjords in the municipality, the outer part of Arfersiorka, called Sarfaq (the current) on account of the exceptionally strong currents caused in this part of the fiord by the tides sweeping in and out four times a day. This constant movement helps to keep the area ice free and navigable throughout the winter in spite of the lengthy periods of frost, with temperatures often reaching -25° to -30° C, and occasionally even lower. At this time of the year the remainder of the fiord system is frozen over, and the west ice closes the coast line from Sisimut northwards. The currents also contribute to a rather rich supply of wildlife resources in the fjords around Niaqornaarsuk, to which should be added the advantage of ice-free navigation allowing optimal exploitation of the resources all year round.

The inhabitants of Niaqornaarsuq have become identified with this meeting of currents to the extent that the term Sarfarnioq has come to be synonymous with Niagornaarsunioq (meaning 'from Niaqornaarsuk'). They speak a particular variety of the local Greenlandic dialect, and the oldest of the settlement's two football clubs is named after the year of its foundation and the place name 'Sarfaq 62.'

The name of the fiord system, Arfersiorfik means 'the place where the whale was sighted', a whale which according to local legend was sighted very many years ago by the inhabitants of the area, and was chased from the fiord towards the coast. There it struck a blow with its tail, thereby touching the place called Attorpa in Greenlandic, which is now the site of another Kangaatsiaq settlement appropriately named 'Attu.'

The Socioeconomic Structure of the Settlement

Niaqornaarsuk today numbers 66 households with a total of almost 300 inhabitants. A little more than 1/3 are children of school age and younger (<15 years). The number of dogs in the community is nearly twice as large as the number of people. As is only natural, most of the male population makes a living as fishermen and hunters. Most of their spouses are employed in the traditional manner as active wives of fishermen and hunters.

The fishing fleet is made up mainly of small fiberglass boats powered by outboard engines and a few boats made of wood or canvas. There are almost as many boats as there are fishermen and hunters. In addition, a few larger craft are found: 11 small fishing boats built of wood, and five slightly more modern fishing vessels of which the three largest are made of fiberglass. Finally, an old, 19.36 tonne wooden cutter is registered at the settlement which is used for shrimping along the coast. Catches are usually landed at Aasiaat.

Apart from employment by fishing and hunting, employment in the settlement is characterized by a number of permanent public sector jobs, full-time jobs as well as part-time jobs, including employees of the municipality and various branches of the Greenland Home Rule government and of Danish government.

The branch office of the municipal administration is manned by four full-time and one part-time official. In addition, the branch office employs three full-time child care givers, 5-6 part-time domestic helpers, one garbage collector, two parish bailiffs, a married couple to manage the youth club two evenings a week, and a manager of the local power station. In the winter four men supply the settlement with water by dog sled twice a week, and in summer two full-time employees operate the municipal water-supply boat. As the need arises, the branch administration office employs a variable number of temporary labourers. It also operates the local skin hut to which the Home Rule government contributes a small amount for the seasonal employment of three women and one man. Finally, the municipal school employs 12 full-time and seven part-time staff.

The various branches of the Greenland Home Rule government, including its business enterprises, also provide jobs. The KNI (Greenland Trade) branch in the settlement (comprising office, shop, and storerooms) has 12 full-time employees. The number of employees at the local salting house run by the Greenland Home Rule production enterprise

KTU varies a great deal because of seasonal variations in the quantity of raw materials available. On an annual basis, the total amount of work performed clearly makes this the biggest single place of work in the settlement. The church employs one organist, one verger, and two cathecists who officiate at most services and other ecclesiastic functions in the school chapel of the settlement. The Danish Health Service employs two mid-wives at the local nursing station.

Privately owned businesses include small building and plumbing firms each with two full-time employees. Both firms are run by people born outside Greenland. There are also two small tobacconists run as private family businesses. Finally, several of the inhabitants receive some additional income in the form of allowances which they are paid as elected members of Kangaatsiaq municipal council (2 persons), the local settlement council (5 persons), the church council (4 persons) and the school board (3 persons).

Thus, roughly 1/5 of the settlement's adult population is engaged in full-time occupations other than fishing and hunting. A further 1/4 regularly or occasionally work for wages. On the other hand, it is worth noting that virtually every male worker in this rather eclectic mix of wage earners engage, to a greater or smaller extent, in fishing and hunting to cover the needs of their own families.

Beside this division by occupation, the settlement can be viewed as being divided into a network of increasingly comprehensive and mutually overlapping circles of communities centred around each independent individual (cf. Rink 1982). A basic characteristic of social interaction, including all communal situations, is the highly developed mutual independence of individuals. This can be seen as a result of the still extremely traditional pattern of socialization. This independence is perhaps most strongly apparent in the widespread reluctance to interfere with other peoples' affairs and decisions, and a concomitant resentment against meddling in other peoples' business. One aspect of this is a pronounced unwillingness to give and take direct advice, let alone orders. Instead, when the need arises, collective exhortations are preferred, like 'let us do this or that,' or similar persuasive requests such as 'would you perhaps....' To a very considerable extent, the general rule appears to be not to interfere, even in cases where there seems to be very good reasons for doing so. There are of course limits, e.g., when the common good is directly at risk. Whether one acts wisely or foolishly is one's own business, as long as it affects no one else. There is no need to explain one's actions to anyone and one's motives are one's own.

Hunters and fishermen are brought up and become individuals capable of choosing and planning their activities in accordance with external conditions (e.g., the local ecology with its regular cycles of abundance/scarcity and seasonal weather conditions) as well as the more unpredictable variations in the 'stocks' of game and fish, and the fluctuations of prices. Nor must they forget to make realistic assessments of their own abilities, decisions, and needs.

By far the most important source of income is fishing for cod, which can be done nearly all year round, although to a lesser extent during the winter when only non-migratory and Greenland cod are available. The summer cod, which migrate into the fjords every spring mostly in huge numbers, is a very valuable seasonal resource until the autumn when they leave the area. Outside the netting season, cod is caught by one man alone or by a couple of men who jig from small boats, or two or more persons working together from a small fishing boat. It is always considered best to work in pairs when setting and tending nets.

Next to cod in economic importance is Greenland halibut which is also caught by net, but only in the winter. The catch of Greenland halibut varies from year to year, as is also the case with summer cod. Finally, a smaller number of wolffish are caught and sold in the period from July to October. This fish is usually caught by long line, a method often used by one man alone in a boat.

In the subsistence economy sealing is the most important factor. At the same time, this type of hunting calls for more knowledge and skill than most others, and is usually a one-man job. The most important catch consists of ringed seals which are taken primarily in winter and spring either by rifle (from small boats or on the ice) or in nets set under the ice. A fair number of harp seals are caught in the autumn, when they migrate in large herds to the fjords. A small number of hooded seals and bearded seals are also caught, as are, occasionally, a few beluga and narwhal.

Hunting, which is the dominant activity on land, is a very important element of the subsistence economy; caribou meat is a welcome variation in a diet of fish and meat of marine mammals. To many families, caribou meat has become a fixture of festive menus. Caribou is also a good source of income, as the meat can be sold at fairly high prices either in the settlement or on the open air market in the 'big town', Aasiaat. Caribou hunts are conducted to the south, in the area around the Nassuttooq Fiord, and mostly in the autumn from August to October, though in some years also in March by dog sled. The hunters who usually leave together in pairs, and sometimes more, are often members of the same family. Generally, this is for practical or social reasons, as the actual hunt is mostly a one-man affair, although a limited form of cooperation is sometimes practised when driving the animals.

Some hare and grouse are caught for household use, but only on a very modest scale. By contrast, the catch of seabirds and web-footed birds is quite an important source of food, including various species of duck and razorbill, gulls, terns, and especially brunnichs, guillemot, and eider, of which, in winter, some are sold in the open air market in Aasiaat.

Although hunting is an occupation requiring an extraordinary degree of independence, each individual hunter is a member of a solid and deeply connected family community. He obeys and helps to uphold the rules of the community by which he in various ways contributes, and from which he also benefits. This is a primary community, deeply rooted in history, which generally encompasses several supporters and part-time supporters, i.e., hunters as well as family members earning some kind of wage income. Together, they support and provide resources for those working at home and other members of the family. In other words, this is a community which can be seen as a sort of extended family which includes married couples and single adults as well as their children or foster children. They are all tied together by virtue of various degrees of family relationships, comprising two to three generations and distributed over one or more households. The various tasks, rights, and obligations are apportioned fairly traditionally according to sex, age, and family attachment.

To the hunter this means that he cannot completely control how he spends his time nor the use of his tools. In certain situations and at certain times (where he may not really want to) he will, at the very least, feel obliged to cooperate with other members of the family in certain tasks, and/or be kind enough to lend them some of his personal tools and other belongings. But since the family community is largely based on 'give-and-take' in such relations, every member of a family is entitled to expect, at least in principle, that the others will feel equally committed to the community. By participating in the community, each member enjoys the additional advantage of being able to fulfill some of his or her needs. Not only social needs, but also personal needs.

The line between each family member's personal identity and independence in relation to the family community is, to a certain extent, variable and hard to define. It is largely determined by factors such as sex, age, and the actual situation. In addition, it varies from one family to another.

For the hunter, in his capacity of family supporter, drawing the line between his independence and his family membership must at all times be determined by several considerations. On the one hand, it is mostly left to himself to decide how he wishes to contribute to the support of the family, while on the other, he will in certain situations feel rather obliged to cooperate with other family members in certain tasks. Generally speaking, the relationships of each member to the family, as well as between family members, are more or less constantly being reviewed and redefined in accordance with the family's changing composition with respect to age, sex, and number over its lifetime.

As a child one gradually develops independence, so to speak, by being left to look after oneself to an increasing extent, a logical corollary of the fact that one can only instill independence by exercising it. At the same time, the child is gradually entrusted with various family tasks corresponding to its sex and stage of development. This means that the child is, to a large extent, left to accept independent responsibility when it is believed that it is ready and able to accept it. This is a logical consequence of the way in which the upbringing of children is primarily conducted by means of positive feedback, or to put it differently, by praising what is being done right, rather than by censuring what has been done wrong. As the young grow up in this rather undramatic way they are gradually left with, and assume, responsibility for the roles and obligations of adulthood. They are reminded indirectly, rather than directly, whenever they do something wrong or fail to do what is right.

On entering into marriage or permanent union, a person's position in the family is once again redefined, not least because he or she becomes a member of the spouse's family and assumes the corresponding rights and obligations involved, while remaining a member of his or her own family. Furthermore, one becomes a link between the parties involved. Over and above the simple fact of having contributed to the family's growth, parenthood also leads to a change of position within the family. In this way, the 'extended family' forms a basic socioeconomic community. Although some households include only one married couple and their children, and would thus in other connections be described as a nuclear family, several factors make it analytically more convenient to use the 'extended' rather than the 'nuclear family' as the smallest definable socioeconomic system. Within this definition, the isolated nuclear family could be regarded as the narrowest borderline case.

The reason is that most households in the settlement include persons who are related to each other in ways that reach beyond core relations (parent-child), e.g., where several generations dwell under the same roof, where siblings and their children live in the same house, households with foster children, and a host of other combinations. Furthermore, virtually everyone living in what might be termed a nuclear family constellation participate, like everybody else, in family transactions with members of their families living in other households. The transactions correspond in general to those practised in extended families.

Within the extended family, individuals find basic social and financial security. Whether one is short of fishing or hunting tackle, has lost his/her job, is undergoing a marriage breakdown, needs a place to stay, has a child that needs care, or is in need of help in any other way, the extended family is the solution. And, it is of course reciprocal within the given framework. In other words, one is prepared to offer whatever help one can give, when and where it is expected.

This family community is so basic compared with other forms of community because of its finality. Where, when, and if all other communities prove inadequate, or even break up, the individual will always be able to rely on the all-embracing family solidarity, provided, of course, that he is prepared to meet the minimum conditions for membership.

In addition to the continuous cooperation and cohesion within the family, worker solidarity is particularly evident in the pound net fishing season. This method of catching cod and Greenland cod begins in May and lasts throughout the summer. It involves heavy and relatively labour intensive work, as tending a small pound net calls for at least three boats in a coordinated joint effort when setting and raising the nets, and even more so when tending them.

At the same time, this catch produces the biggest cash income in the shortest time which, from the point of view of the cash economy, makes it easily the most important and also the most labour-intensive activity in the settlement. This affects not only the pound net fishermen themselves, but more or less the entire settlement, because at the height of the season everyone capable of wielding a cleaver, including older schoolchildren and senior citizens, are busy helping to handle the catch.

Another activity in which members of the extended family are engaged together, and thus make the family community visible, is the catching of capelin, a small fish which invade the fjords in large shoals to spawn during the month of June. Nearly all age groups participate in this catch, either by collecting the fish at the water's edge with landing nets, or by putting them out to dry on the rocks. The dried capelins are an important part of the winter stores, useful food for man and dog.

A third recurrent joint activity takes place when the families leave for summer camp near one of the many streams in order to fish and smoke Arctic char. This too is an activity in which virtually all age groups take part, particularly towards the end of the summer when there are berries to be picked at the same time.

At the beginning of the summer, many families make one or more excursions to collect birds' eggs, a form of family cooperation which may be said to be more important socially than economically.

The fact that the family constitutes the basic community should not be taken to mean that it is invariably, and in every situation, given the highest priority. For a number of reasons the individual may have social and other motives for entering into other communal arrangements than that of the family. And the very act of having, in certain situations, to set priorities relative to the family and other communities gives the individual an opportunity to demonstrate independence *vis-à-vis* the family as well as with respect to the other communities concerned.

A member of a family community is also part of an interfamilial community through his/her own marriage and those of his/her parents or siblings. This entails a certain measure of cooperation and exchange of catches and services. It goes without saying that one also finds groups of friends within which, as a rule, similar patterns of cooperation and exchange exist, not least because in many cases they coincide to a certain extent with the interfamilial community. However, this does not prevent friendships from functioning and meeting needs similar to those of friendships found elsewhere in the world.

In regards to more extensive and decidedly local social communities, the establishment of 'Sarfaq 62' and the younger 'NBK 89' football clubs is worth mentioning. Externally, both represent the settlement, whereas internally they inevitably produce a certain social division, at least at the level of sports. Apart from arranging and participating in matches against each other and outside clubs, both clubs also organize many social events like bingo games and dances in the local settlement centre.

Similar and other kinds of social events are also held by the local branches of national, hierarchically-structured, larger groups like the trade unions, the temperance association, and the political parties. Through such events, the local branches try to integrate their members and look after their interests within the context of the settlement, while at the same time incorporating them into the larger community structure and external world. Locally engaged employees of the public enterprises and institutions are similarly participants in an integrated locally functioning working community, which is related outwardly and upwardly as a subsidiary branch of a more comprehensive hierarchical system.

The settlement community, which is more extensive and correspondingly less integrated, is evident in everyday life simply by the fact that the inhabitants of Niaqornaarsuq live in the same clearly defined geographical area. It is a community which is isolated from other inhabited areas, and which is occasionally manifested overtly through recurrent events, e.g., the annual football cup finals and the joint dog-sled racing finals.

Regardless of their intensity, all the communities share the quality of being present. In other words, they are based upon and mutually linked by the fact that everyone knows everyone else in the settlement, either directly or indirectly. This means that everyone knows where the others live and with whom they have family relations. Furthermore, communication is mostly direct, personal and reciprocal, and only a few have more influence than everybody else.

On the other hand, the various groups within the settlement are also parts of more complex, dominant, and consequently noticeably more abstract communities. Although most of the Niaqornaarsumiut have relatives and friends or acquaintances living in Kangaatsiaq or other settlements in the municipality, it is impossible to know all the 1,300 inhabitants of the municipality, and hardly even all the employees of the local administration. Communication at this level is more often indirect, by telephone or by letter, and is frequently based on the professional roles and statuses of people in the interaction rather than on their personal relationship.

At higher levels in the hierarchy this impersonal and rather abstract form of communication becomes increasingly dominant. It also grows more asymmetric in that it tends to turn into a one way form of communication directed downwards with increasing structural and personal distance. The publication of collectively addressed, impersonal general regulations and circulars would be an example of communication at this level.

In the context of the settlements, such general orders and regulations received from outside easily come to be seen as irrelevant, if not incomprehensible. By accepting their jobs, the local heads of administration have placed themselves in a sometimes difficult intermediate position by being, on the one hand, physically present and members of the local communities while, on the other hand, having to represent and from time to time enforce directives issued by far-away bodies.

This problem is aggravated by the fact that only a limited number of people in the settlement are more or less attuned to hierarchically structured methods of work management. In many different connections an obviously engrained form of collective division of labour is employed. No leader or manager is explicitly appointed, and each individual participant just slips in and finds a suitable place in the process. This would apply for instance to a festival committee where those involved will volunteer to undertake various tasks individually or even just do them together. It would also apply to the launching and landing of larger vessels, at which the owner will procure and install the necessary tackle, etc., but where passers-by and other more or less accidentally present assist at their own volition and share in the work as their inclination and strength permit. Finally, it would apply to an occasion when a collective whale hunt is initiated, and the participants 'join in' without having to be asked to crew on boats and contribute whatever fuel and time and hunting effort they can muster.

A Collective Whale Hunt[1]

One day my "brother-in-law," Qaqi, came bursting in shouting that now was the time, if I wished to join a whale hunt. Qaqi does not own a boat, but since we belong to the same family he could of course borrow mine or, as in this case, we could go out together. The weather was calm, there was hardly a ripple in the fiord ... ideal weather for a whale hunt.

---

1. Authors' note: The names of all persons in this report are pseudonyms. Editoral note: The authors integrate personal diary entries with narrative to describe a collective whale hunt in this section, and whaling by cutter in the following one.

Good weather is required for initiating a whale hunt. Who then starts a whale hunt? Who takes the initiative? There is no single person with any formal or informal authority to do so. A collective whale hunt is never planned in advance. When the right conditions, the need, and the mood are all present, a whale hunt may ensue.

There are always some small boats on the local fishing banks near the settlement. Almost daily the whales can be observed going in or out of the fiord looking for food. There are many big and powerful currents in the fiord system. This provides the local population with excellent fishing and hunting opportunities. Evidently, it is also a good area for whales.

A whale hunt may, for example, get under way when the fishermen on the banks observe a suitable whale. If the weather is calm, and enough fishermen are in the mood for a whale hunt, the fishing lines are pulled in and a general pursuit of the whale begins. Even when fishing for cod, the men usually bring at least one gun. The first shots at the whale can, of course, be heard in the settlement, leaving none in doubt that a whale hunt is getting under way. Several gunshots fired almost simultaneously can only mean a whale hunt; sealing and bird shooting are individual affairs.

If people on land do not feel like joining in a whale hunt, no boats leave to participate in the chase. Whether one happens to be in the mood for whale hunting or not is an important factor. If not enough boats join in during the first hour or so, the hunt is abandoned and the fishermen return to their relatively mundane fishing job. But if, on the other hand, there is a general mood in favour of a whale hunt there is a great bustle of activity in the settlement as people scurry about, finding, buying and readying. Before long a fleet of small boats sets course towards the sound of the guns.

Whale hunts often take place in autumn after the caribou hunts are over, and before the late autumn southward migration of seals. During that period bad weather may keep fishermen and hunters land bound for days on end. So this is a time when landing a whale may serve as a food buffer for people and, not least, their dogs. This 'buffer' function is particularly important in settlements where access to and consumption of European food is much lower than in the towns.

In Greenlandic households a sharp distinction is made between 'Danish food' (*qallunaamineq*) and 'Greenlandic food' (*kallaalimineq*). Semantically, the category *kallaalimineq* primarily refers to seal meat and whale meat and, to a lesser degree, char and caribou meat. 'Greenlandic cooking' is synonymous with the boiling of meat. Eating habits are determined from what the person eats and how the food is prepared. When someone is identified as eating Greenlandic food it means that he or she not only is eating fish or meat of local origin, but that it is not fried and that spices are not used with the exception of pepper or mustard after the meat has been prepared. If imported meat is consumed at everyday meals, the person's eating habits are marked as 'Danish' (Petersen 1985).

This distinction between Danish food and Greenland food is far more significant than a merely functional distinction referring to the origin of the food. Eating Greenlandic food is of great symbolic weight in determining whether a person is a true Greenlander, and in this connection whale meat, and in particular *muktuk* is very important.

> Well, I put on some warm clothes. Although it was summer still and the sun was in the sky most of the day and night, sitting in a small boat on a Greenland fiord for a whole day can be a cold affair. Qaqi placed his guns, the 6.09, for the whale hunt and the shotgun to be used in case a likely bird should appear, as well as his harpoon in my boat. We borrowed a float from Aleq.

The harpoon is indispensable in collective whaling. It is used to attach floats to the whale. This makes the hunt much quicker and more effective. Qaqi's harpoon has been made by his father-in-law, Aleq, who is one of the few that still master the old crafts. For years Aleq has taught hunting and fishing at the settlement school. The front shaft of Qaqi's harpoon is made of bone. The whale's lower jawbone is immensely strong and ideal for toolmaking. Harpoons made of traditional materials are best, but a great deal of skill and time is needed to produce them. Today, most harpoons are equipped with brass front shafts, or even just a screwdriver.

> On the way out we stopped at the tank. While I bought 50 litres of petrol in cans, Qaqi ran to the shop to buy 200 bullets. Then we were off, heading for the other small boats. When we reached them there were already nine other boats. Including ourselves and another fresh arrival, the little fleet totaled 11 boats. There are two men in each boat, one steering the outboard engine and one standing at the bow doing the shooting. Needless to say, shooting is most exciting, so in many boats the two change place(s) several times during a whale hunt.

A formalized sign language is employed in sealing and whaling. The man doing the shooting uses it to direct the helmsman. With one hand along the side of the boat he directs the speed of the boat. Waving the hand obliquely downward towards the rear means 'reduce speed', waving it obliquely upward and forward means 'increase speed.' The helmsman really has to show his skills, accelerating and decelerating just slowly enough not to upset the balance of his standing mate.

Furthermore, by raising and lowering his outstretched arm, the standing man can indicate a new fixed point and a change in course. The signs are not specifically local, and there are in fact reasons to believe that they have been adopted after the introduction of the outboard motor. At any rate, the infernal noise produced by the engine effectively prevents any verbal communication between the two persons in the boat.

There is another sign language relating to seals and whales. It consists of only two signs, both of which are widely used and known to be very old. Clenching a fist with the palm of the hand towards oneself, the forearm held vertically in front of oneself and moved up and down a few times means 'seal.' The symbolism is quite clear: the head of the seal rising vertically out of the water and going down again. The other sign means 'whale': bending one hand slightly at the wrist, keeping the fingers together and slightly curved as an extension of the curve of the wrist, and resting the thumb against the index finger. This sign is as clear as the other: that is how a whale looks when surfacing to breathe. During a collective whale hunt seals may appear between the boats. Then it is very useful to be able to communicate quickly to the other boats that it is 'merely' a seal and of no interest just then.

> On reaching the others we go alongside one of the boats. The most essential information about the species of whale and its position is given. To begin with we were just waiting for the whale to appear. When it surfaces shots are fired at it, and the boats race to the position where it surfaced. The first shots are fired when the whale is still well out of range.

At the initial stage some of the hunters may choose to land and climb one or more observation posts, rocks from which there is an unobstructed view, mostly over two arms of the fiord. From there the whale is observed, its species and course discussed. This does not mean that a detailed hunting strategy is worked out; that is neither necessary nor indeed possible since no one knows where the hunt will end. There is, however, one basic strategy: the whale must be prevented from leaving the fiord, because, if it reaches the mouth, the heavy swell coming in from Davis Strait will make it impossible to hit the whale while standing in a small boat and the hunt will have to be called off.

A collective whale hunt can be divided into two stages. In the first stage, it is a question of finding the whale and tiring it out by shooting at it whenever it surfaces to breathe. This gradually brings the hunters closer to their prey. Few bullets, if any, actually hit the whale during this first stage. Some whales manage to escape at this stage, long before they have received any hits at all. They escape either because the hunters lose sight of them or by swimming out of the fiord.

During this first stage, the whale is often submerged for a long time, and in between the hunters have no clear idea where it may be. If the hunters feel they have lost touch with the whale, they place their boats in a fan-shaped formation moving away slowly from the centre. No one is directing individual boats to their position in the fan. There is little need for any conversation; everybody is perfectly aware that contact with the whale has been lost. Each boat finds its place in the fan and searches its own field. There is no formal or informal leader of a whale hunt. Every now and again someone may give an 'order' to sail in such and such direction. An instruction like that is usually complied with. Whoever gives an order either tells his reason for doing so, or is known to have a concrete reason for ordering others about. Orders are usually in the form of exhortations. A valid reason for giving orders is, for example, to have sighted the whale or at least to have a well-founded idea of where it may be.

At this juncture the hunters' familiarity with the way animals think is of crucial importance. Each hunter coolly tries to calculate where the whale is likely to surface next. His knowledge of the depth of the sea, its currents, etc., in combination with the distance to coast, the movement of the tide, the direction in which the whale submerged and other considerations give the experienced hunter a shrewd hint of where the whale is going.

One of the indications of the whale's position to the inexperienced eye are the nearly invisible ripples on the surface of the sea produced by the whale pressing water through the baleen. A hunter observing such ripples immediately alerts the others by arm signals, the sign meaning 'whale', followed by pointing to the place where the whale was last sighted. Everyone then speeds along in that direction.

This first stage of a whale hunt may last several hours. The number of escaped whales vary in inverse proportion to the length of the pursuit. Sooner or later the whale begins to surface more frequently, and it no longer dives very deep. This is where the second stage of the hunt begins, the kill. By now the hunters are in eye contact with the whale most of the time. The boats fall into line behind the whale. Whenever it surfaces to breathe it is being shot at from all boats.

As noted above, the fiord must be calm before a whale hunt can start. This is because the less the boats rock, the better the hunters are able to find their mark. Most hunters are first rate shots. It takes years of training to be able to hit a seal between the eyes with the first shot fired from a small boat. At this stage of the whale hunt a hit rate of some 70-80% is achieved.

It is no accident that stone throwing is the favourite game of the settlement's boys. Child's play results in an adult's skill. Many Greenland games include an element of precision throwing. From a tender age onward Greenland boys are trained to become good shots. The boys are very inventive. They will cut off the spout of a plastic container. Around the salting house a discarded rubber glove can easily be found. The thumb of the glove is cut off, rolled over the spout and firmly tied to it. In Greenlandic this kind of 'thumb sling' is called *kullukoq*, meaning 'cast-off thumb.' Pebbles are placed at the bottom of the elastic rubber thumb, while the spout is held between the thumb and index finger of the other hand. The boys achieve remarkable standards of marksmanship with their slings. They can often be seen on the landing bridge shooting at flying seagulls which they attract by throwing fish offal in the water.

> As at all collective whale hunts, we had to try to harpoon the whale. As a special feature of this particular whale hunt, one of the settlement's worthy hunters was allowed to let one of his sons be the first to try. It was not only the first try at harpooning this whale, but also the young hunter's first harpooning of a whale.

Pernatoq is Greenlandic for something happening for the first time. This 'first time' concept is prevalent all over Greenland, and indeed in all Inuit regions. Time and care is taken with a first time harpooning like this, even at the risk of the whale slipping away. The first time performance by a young hunter of some element of his professional repertoire of skills confers some special social obligations, for instance more gifts of meat. This usage still applies, for instance, in connection with collective whale hunts.

Formerly, a young hunter would also have to undergo certain rites in connection with his first catch. The first time Aleq caught an *allatooq*, a 2-3 year old harp seal, he had to eat a little of all parts of the seal, without

Inuit And Whales at Sarfaq (Greenland)

using his hands! These rites are no longer practised, but they are still known and talked about.

Similar rites used to be connected with a first time whale catch, particularly of a small cetacean. A custom unconnected with the 'first time' concept, but having something to do with whales is worth mentioning. If someone is out sailing, without being on a whale hunt, and is bothered by a whale that feels like playing around with the boat, then all one has to do is to drop a knife into the water and the whale will disappear. This custom is still practised.

The whale is regarded with respect by the hunters. It can be dangerous if one gets in its way. A whale can easily capsize a small boat. More than once during the hunt all the boats dashed away from the whale. This happened when the whale altered course on the surface, for instance, when it was swimming straight towards an island and suddenly realized that its way was blocked.

Collisions between boats do occur and sometimes one of the hunters falls overboard. That is the end of the hunt as far as he is concerned, at least until he has been back home to get a change of dry clothes. In order to avoid falling into the sea, some hunters will tie themselves to the boat.

> The whale was harpooned with two floats, a device which makes the second stage of the hunt far more effective. The floats break the surface of the sea before the whale appears.

This means that the hunters are ready to shoot as soon as the whale surfaces. By sailing up in front of the whale, the hunters try to guide its course. The aim is to get the whale into shallow water just before it dies. This does not always meet with success. The hunters had chosen a small inlet where it would be convenient to flense it. Some boats were sent up ahead of the whale, but they failed to alter its course. Instead, the boats had to make a hasty maneuver to get out of the way of the whale forging ahead. Shortly after it died.

A collective whale hunt is not planned beforehand. This may mean that there are not enough large floats on hand. That is precisely what happened in this case. As soon as the whale died it began to sink. The only link ensuring contact with the whale were the two harpoon ends stuck in the whale, to which the floats were attached.

The whale sank to a depth of some 80 metres. During the hectic minutes when it was going down, the hunters had been quick to lengthen the lines to the floats. I confess that at the time I had written off the whale, but the hunters sent for a small fishing boat to which the line was fastened. First by muscle power and later by engine power they managed to drag the whale towards the shore. Saving the whale from being lost depended solely on the two little harpoon ends embedded in its skin. At high tide it was hauled close to land and as soon as the tide started going out, the hunters started flensing the whale.

Whaling by Cutter

> On a day in October, Masik and I were out hunting seal. As the southward migration was late we had sailed to the mouth of the fjord to try our luck, but to no avail. As it began getting uncomfortably cold we started gloomily for home. On the way we met a cutter hunting a whale, and immediately decided to join the small boats assisting the cutter.

Unlike collective whale hunts, whaling by cutter is usually planned in advance. A cutter may be granted a 'quota' whale, and will then set out on a planned whale hunt. One of the conditions for being granted a 'quota' whale is that the cutter must be equipped with a harpoon gun. The first stage of a cutter whale hunt does not differ materially from the first stage of a collective hunt. The whale must be found and tired out. In a cutter whale hunt, the small boats employ the same fan order when searching for the whale.

The second stage is shorter than in collective whale hunting. As soon as the cutter gets within range and hits the whale with its harpoon gun the hunt is over. The whale is then towed to the nearest suitable flensing ground.

Flensing a Whale

> We put into a calm inlet. The whale was pulled up at high tide. There was a full moon providing enough light to work, although it was night. As it was spring tide the whale came well up on land. There was no wind, and the water was calm and easy to work in — ideal conditions for flensing. It was a beautiful night over a small group of men flensing a whale in the middle of nowhere. The household would have meat again, though not until the morrow for that is when the meat will be shared out. Everybody should be granted the happiness, the delight of standing on a rock on a calm Greenland autumn night with a full moon and a light frost, after a successful whale hunt. Just to stand there listening to the immense silence — the silence encompassing all sounds.

Not all places along the coast make equally good flensing grounds. Knowing the good places makes the flensing far more efficient. There are few places as suitable as the one near Sisimiut, an old-established flensing ground which has the special advantage that the tidal current helps to turn the whale when one side has been cut clean. The flensing as such is basically the same irrespective of whether it is a whale caught by a collective or a cutter hunt. The tail is highest up onshore, the head facing the sea.

As soon as the whale has been safely brought on land, the hunters bring out their pocket knives to cut a slice of the particularly tasty *muktuk* found on the tail fins. *Muktuk* is whale skin with the underlying layer of blubber. It is considered a real treat. Fresh *muktuk* is eaten raw. It has a very high vitamin-C content. With large kitchen knives, saws and axes some of the hunters cut out first the layers of blubber and then the meat. It is cut into the largest possible pieces that can be carried onshore across the tidal zone. This job is done by some of the others. Others again distribute the pieces among piles representing shares of the catch. First the immediately accessible parts of the whale are flensed. At high tide the whale is turned, and at the next low tide the remainder of the whale is flensed.

Distributing the Catch After a Collective Whale Hunt

As soon as the first side of the whale had been flensed, everything was divided into 17 shares of the catch. Usually the number of shares is based on each boat getting one share. In other words, boats where the two men onboard come from different households and boats manned by two from the same household get the same amount, one share of the catch.
A share of the catch consists of three parts:

- *Neqi*: Meat in general. No distinction is made between different cuts of meat;
- *Orsoq*: Ordinary fat, including, of course, the attached muktuk; and,
- *Qiporaq*: The blubber making up the belly flesh. This blubber is considered the best, partly because it is more tasty, and partly because it is more tender.

Considerable effort is made to ensure the fairest possible distribution. To begin with, a long time is spent arranging meat and blubber into piles. The hunters will then discuss whether there is too much in one pile or perhaps not enough in another. When everybody is agreed that the piles are all as near equal as possible, the actual distribution begins. First, an indicator and a *naalagaa* (person to be listened to or obeyed) are appointed. These individuals are responsible for distributing the 51 (3 x 17) shares of the catch. The hunters, each representing one boat line up facing towards land. The 51 piles are placed at the water's edge. The *naalagaa* takes up a position between the hunters and the shares, with his back turned towards the latter.

The meat shares are distributed first. The indicator points at a pile asking: 'Who is to have this share?' Without knowing which share, the *naalagaa* replies by naming one of the hunters in the line. The hunter named gets that particular share. When all the meat has been distributed, the *orsoq* and *qiporaq* are distributed in the same way. Today, the distribution of a whale caught by a collective whale hunt is by quantity and by quality. Each hunter receives the same quantity, but it is considered as important that the quality as well should be as near equal as possible.

Previously, small cetaceans in particular were distributed by quality alone, depending on the participation of each hunter (cf. e.g., Hertz, 1977: 92-95). Strangely enough, the number '5' is frequently found in traditional Greenland thinking. Hertz says that the first five participants in a catch get preferential shares. At the local dog sled races the first five win medals. They are called *angupput*. Those coming in '6th' place or lower are called *angutinngituulaarput*. In Western thinking it is, of course, the number '3' that holds a special position.

The actual distribution of shares to the boats participating is, however, merely the first step in the distribution of a collectively caught whale. There are three levels of distribution:

1: distribution of the share of the catch as described above;
2: distribution of each share among the hunters in each boat;
3: gifts of meat and various ways of preparing and storing the catch (freezer, *nikkut*, etc.)

At the second level of distribution, the allocation of individual shares of the catch between the hunters that have sailed in the same boat, there are also fixed rules. Each person gets a share, but the boat gets one as well. If the owner of the boat has paid the petrol used himself, the boat's share goes to him. If the men in the boat have shared the cost of petrol, the owner gets 2/3 of the boat's share, the other hunter gets 1/3. If the owner of the boat has incurred no expense, the boat's share is divided equally among the two. These rules do not apply only to whaling. They are part of a large system of socially accepted rules of distribution of catch when boats or other equipment have been borrowed. Depending on whether or not the owner himself has taken part in the hunting/fishing, as well as on whether the borrower has paid the costs incurred in connection with the hunting/fishing, the borrower settles with the lender according to fairly fixed consensual rates. These rates are in proportion to the size of the vessel used, from 20-30% for the owner of a small boat to 55% in the case of large cutters. In cases involving large boats, money is used to maintain reciprocity, but payment for borrowing a small boat is usually in kind.

When the meat and blubber from the whale have been distributed to the households directly involved in the hunt, a further allocation, at a third level of distribution, is made. This is to ensure, by means of gifts of meat, that households not involved in a given whale hunt receive some benefits as well. Nowadays, whale meat gift distribution takes place primarily within each hunter's family, i.e., to his parents, siblings, children with households of their own and a few cousins and others with whom he has some special connection (e.g., name relations). All households that have received some of the catch practise some form of preparation or preservation of the meat and blubber (freezer, *nikkut,* etc.).

Most of the *muktuk* is eaten during the first few days following the catch. Fresh *muktuk* for breakfast is a good way of starting the day. Of the blubber brought home, a good deal is dried and stored as dog food. It is of great nutritional value in winter because of its high energy content. Some fat, primarily *qiporaq,* is stored as human food. It is eaten raw or boiled with the meat of the day. In Greenland cuisine, *qiporaq* plays the same role as potato and bread in the European diet.

The whale caught in the hunt described above was distributed locally only. Now that the freezer has become widely used, some of the meat was frozen. Dried whale meat, however, is particularly popular. Before drying, it is cut into smaller pieces. In Northwest Greenland it is cut into long strips, whereas in Southwest Greenland it is cut in flat slices.

Half the population of Greenland live in municipalities where the keeping of sled dogs is banned. In south Greenland the absence of winter ice means that there is no practical use for sled dogs. However, they are indispensable in the rest of Greenland. In winter the dog sled is often the only means of transport hunters use to get to their hunting grounds. In settlements without municipal water supply arrangements, a family's own dog sled and team is often the only way to fetch water, either from a lake or from an iceberg.

A household with active hunters may keep up to 30-40 dogs, although half this number is usual. There must be some surplus dogs as some dogs may be injured, while others may have young pups and for that reason cannot be used for some time. Providing one's dogs with food involves a lot of work. A whale is a good source of food. Whale blubber is very nutritious, and in winter is an excellent supplement to the dogs' staple diet which consists mostly of fish.

The carcass of a flensed whale is free for all. No individual or group of hunters have any rights of access or ownership over a carcass once the catch shares have been cut out and distributed. As long as the whale is being flensed the place is called *pillattarfik* (flensing ground). When the flensing is finished is it termed *pilattarfikoq* (former flensing ground). There is more to this than a mere change of name, because when the place is called *pilattarfikoq* everyone may take as much as they need. During the first days after a flensing, many hunters go out to pick meat and blubber for their dogs.

During the big *sassat* in the Disko Bay in the winter of 1989-1990 many dog sleds came from far away to fetch meat and the much coveted beluga *muktuk*. All along the west coast of Greenland they benefited from this *sassat*. Some of the *muktuk* was sold to the KTU which distributed it to the rest of the country. But a lot of frozen parcels were sent to family members living in other parts of Greenland (e.g., children away from home studying, etc.). Everyone was given a share of the household's catch.

Distribution of a Whale Caught During a Cutter Hunt

The principles applying to the distribution of a catch resulting from a cutter hunt differ from those applying to the distribution of a whale caught by a collective hunt. In a cutter hunt fewer people are involved in catching a whale, which is usually larger. From a legal point of view, the entire catch belongs to the cutter and the meat is usually sold in the port of the owner. The general principle of distribution in the case of a whale caught by a cutter hunt is a quantitative one, and there are only two levels of distribution:

1: the crew's wages, plus a small amount for the families; and,
2: sale in the open air markets in the towns.

The crew is paid mainly in accordance with percentages fixed by their terms of employment. A certain percentage of the proceeds of selling the meat is set aside to cover the running of the cutter. Then the skipper and his crew receive their pay as stipulated in the relevant agreement. This does not, of course, mean that each crew member cannot take a certain amount of meat for private consumption, but it is a much smaller quantity than in the case of a collectively caught whale. A small part of the meat received by crew members is distributed as meat-gifts, though not nearly to the same extent as in collective catches. By far, the largest volume of the meat is sold in towns, where most of the population has no opportunity to participate in whale hunts. This sale takes place in the open air market, a place where local hunters can sell their catches direct to the consumers.

The prices per kilogram differ from product to product. There is consequently a clear connection between price and quantity. As for quality of meat, there is also a clear relationship with price. Insofar as quality is taken to refer to different kinds of animals, the price of char is different from that of cod, and so is the price of seal different from that of whale. In Western concepts of price differentiation, the price per kilogram also depends very much on the cut and the part of the animal from which the meat comes. In this concept of price differentiation one finds an undisputed hierarchy, for instance between a tenderloin and spare ribs. This Western concept of price differentiation is universally used in Greenland shops, but with open air market sales a Greenland concept of price differentiation is in force, at least partially. The idea behind this Greenland concept is that each individual may have their own preferences concerning meat, and hence it is impossible to set up an absolute sale price and quality for each different cut of meat from an animal. An expression of this concept is the fact that leg of caribou is sold at the same price as ribs. Similarly uniform prices are, of course, charged in the case of seal and whale meat. To an even larger extent, uniform prices are charged in the settlements where the recommended prices issued by the westernized central administration are widely disregarded.

Occasionally a whale is caught jointly by a cutter and a fleet of small local boats. In such cases a portion of the catch falls to the small boats partaking in the hunt. It is up to the skipper of the cutter to decide how much the small boats receive as their joint share. In general, the small boat hunters are well satisfied with the quantity allotted to them.

In the case of the whale caught by the combined cutter and collective small boat hunt described above, we actually had to ask for a share of the *qiporaq*, one of the three groups into which the catch is divided. Following the division of the catch between cutter and small boats, the detailed distribution is carried out as previously described.

Households Involved in Whaling Activities

In 1989 approximately half of the 66 households at Niaqornaarsuq were directly involved in whaling. As noted above, there is no fixed pattern nor any plan determining who takes part or when. Some are more enthusiastic than others and manage to participate in a collective whale hunt nearly every season, whereas others join in only occasionally. Some households are never, or almost never represented, either because they do not include any adult males, or because the men are physically unable to participate because of old age or for other reasons. Additionally, in some cases some do not take part for more specific and personal reasons. There are others who do not participate, or at least only very rarely, because they do not own a sufficiently fast boat, have no opportunity to sail with somebody else, or are unable to borrow the necessary equipment at the right moment.

Of the more than 30 households stating that they have participated in a collective whale hunt during 1989, 19, or nearly 2/3 agreed to take part in a survey. The 19 households comprise a total of 107 members, of which 38 are men and 26 are women over the age of 14, and 22 boys and 21 girls — altogether just over 1/3 of the population of the settlement ranging 65 years in age. Twenty-eight of the men and four of the boys state that they have taken part in a whale hunt, which means that in some cases two, three, and four members of some households have at one time or another participated in whaling.

On nearly all the occasions stated, participation has been by small boats with outboard engines, mostly manned by two men, but in some instances by three persons or only one. One of the respondents took part on board a small fishing boat. A few whaled with boats owned by friends or relatives outside the household, but a large majority had sailed either in their own boat or in one belonging to some member of one's household. The number of boats in a hunt varied between 12 and 22.

Utilization, Distribution, and Storage of Whale Products

According to the respondents, after each successful hunt the whale is divided in the customary way into equal shares, one for each boat. It is also stated that in every case the entire catch had been distributed. Fifteen of the households report having given away some of their share as meat-gifts; four say they gave 'quite a lot', six gave 'a good deal', two 'a little', and the remaining three merely say that they have gave 'some' of the catch away. In the case of five of the 15 households, meat-gifts were presented to relatives only, whereas the remaining 10 had made presents of meat to both friends and relatives. One single household stated that over and above giving meat-gifts to relatives, some of the catch had been sold.

As indicated by the survey, a small amount of whale meat from local catches is occasionally sold but within and outside the settlement. However, the quantity sold is insignificant compared with that given as meat-gifts, not to mention what is consumed within the households themselves (the income derived from sales of meat will only in rare cases exceed the expenditures on petrol, etc., incurred on a hunt). In addition to this kind of sale, and as with other natural produce, whale meat is sometimes used in barter. As in the case of the exchange of meat-gifts, the parties involved tend to regard barter much more as a confirmation of their mutual relationship than a mere exchange of goods.

An ordinary sales transaction is regarded as formally settled by the exchange of a given commodity for money. In principle, the relationship between buyer and seller may be regarded as being complete the moment the transaction is finished or, in other words, to the extent that the parties no longer owe each other anything. The exchange of meat-gifts, however, involves a symbolic manifestation of the parties' relationship with each other. The regular exchanges in fact represent a continuous definition and consolidation of mutual relations, rather than a repayment of debts to each other.

In a reciprocal relationship, the parties will no doubt eventually give roughly the same amount to each other, but the balance remains more or less open, and so the relations are carried on indefinitely. In the case of a more unequal relationship, as for instance between a skillful and a less skillful hunter, one of the parties will usually have more meat to give than the other, and so the exchange will reflect the naturally occurring unequal relationship in this particular respect, though not necessarily in all respects.

In a similar way, barter is not so much a question of whether a given quantity of meat is of the same value as a given piece of wood, or whether a certain service has been rewarded with a sufficient amount of meat. It is to a much higher degree a question of developing and maintaining friendly, complementary relationships by an on-going, open-ended exchange of goods and services.

Regarding the part of the catch retained by each household for its own use, 16 households state that some of it was used for dog food in quantities varying from 'a little' in 10 households to 'a good deal' in five households. Only one household had used 'quite a lot,' and the remaining three households report that they did not use any at all.

Obviously, the least attractive parts of the meat are given to the dogs, i.e., parts containing sinews and cartilage, pierced by rifle shots, and scraps in general. But whale meat, and in particular blubber, is considered excellent food for dogs because of its 'substantiality' (i.e., its concentrated nutritional value). Those who have managed to keep some of it in store, often feed it to the dogs before embarking on long dog sled journeys.

As a draught animal the sled dog remains an indispensable and important feature of life in the settlements. Dogs are used partly in connection with hunting activities in winter, such as *uuttoq* hunting (the shooting of seals sunning on the ice), tending of seal nets, shooting grouse, hare and other small game, and winter hunting of caribou which usually takes place two or three days' journey from the settlement. They are also used partly for carrying water, as in winter the settlement has to fetch its water supply from a lake 2-3 km away.

All 19 households state that some of the whale had been put into storage for use by their own members, varying from 'a good deal' in 10 households to 'most of it' in the remaining households. The meat is kept either dried or frozen. It was estimated to last 3-4 months in four of the households, about six months in 11 households, and up to a year in the remaining four households. It is widely believed that dried or frozen whale meat, like the meat of caribou, keeps particularly well for long periods because it contains very little fat, and hence does not tend to go rancid in contrast to, for instance, seal meat.

There are evidently no set norms saying how much or what share of the catch it is suitable to give away as gifts. However, up to a point it must quite obviously depend on the amount of meat acquired by the household, compared with how much the household expects to need for its own consumption, and the number of gifts one would feel obliged to give. Of the four households stating that they had not used any of the whale meat as gifts, the largest (with 12 members) stated that, apart from using some of it as dog food, it has set most of it aside as a reserve expected to last about six months. Two other households with six members each have also put most of their share in store expected to last 3-4 months and six months, respectively. The last household with only three members has used some for the dogs, and most of its share as a reserve expected to last about a year.

The two households that had given away 'only a little' state that they have put most of their share in storage expected to last three to four months and six months for six and nine members, respectively. As a comparison, two of the five households stating that they have given 'a good deal' away have been able to set aside reserves for one year's consumption for two and four persons. And the three other households with four, five, and seven members report that they have been able to keep most of their share in storage, expecting it to last between three to four months and six months. In spite of the quite considerable differences in the amount of meat distributed as gifts by individual households, the general tendency appears to be that gifts are given to the extent that it is felt there is some to spare, that is, in proportion to how the household's own perceived needs have been covered.

Finally, 17 households state that they have put aside some of the *muktuk* for festive occasions. As in other parts of Greenland, *muktuk* occupies a special position in people's diet and in their minds. Many families keep some frozen *muktuk* for use on very special occasions like important birthdays, anniversaries, weddings etc. *Muktuk* is usually served, by itself or as part of a special treat, late at night and marks the culmination of the festivities.

Svend E. Larsen and Klaus G. Hansen

Conclusion

A collective whale hunt is at one and the same time an integrated and an integrating feature of the economic, social, and cultural life of the settlement. Fluctuations of climate and seasonal variations in different animal populations make it extremely difficult, if not impossible, to survive by means of only one or a few resources. In the Arctic, hunters have always had to base their survival on a number of different resources available in variable quantities at different times. The traditional, locally based subsistence economy has gradually come to be linked with the now universally prevalent cash economy in a familiar complementary relationship, where local natural resources are exchanged with money that can, in time, be used to buy equipment for the 'harvesting' of natural resources.

The access to imported resources by means of income obtained by trading or working for wages has obviously served to increase the local prospects of survival, partly because resources have increased, and partly owing to the fact that this has made the exploitation of local resources more effective. In spite of this, the subsistence economy remains a very considerable part of the basis of survival for settlements like Niaqornaarsuk. Without that basis it would hardly be feasible to maintain the existence of the settlement. In this connection whales are only one of the wildlife resources of the area. Clearly, the economy of the settlement would not collapse completely if its inhabitants were to be barred from making use of just one out of several available resources. However, if one were to start eliminating one resource after another, one would be narrowing the already limited and precariously balanced set of options, and hence the entire local basis of existence. This is a circumstance which has recently been placed in proper perspective since the sale of seal skins has been made unprofitable. Whaling itself does, however, represent an important factor in maintaining the ecological balance, and the hunters themselves will obviously be interested in keeping their activities at a level that will permit them to continue to depend on all available options.

As a complementary resource, whales are thus an integrated aspect of the economic and social life of the settlement. As a source of food, whales serve as an economically and socially integrating function, especially in view of the quite considerable extent to which whale meat is distributed as meat-gifts.

Collective whale hunting is based directly on the communal life of the settlement, regarding the availability in a given situation of enough men, boats, etc. At the same time, it is an expression of community by virtue of the collective character of hunting and distributing the whale. In other words, whaling as a collective enterprise is an integrated aspect of social life in the settlement, while being itself an instrument of integration by promoting social coherence, partly due to the activity itself and the interplay between the participants, and partly by means of the subsequent distribution of the catch within and between various groups of friends and families.

Rooted as it is in ancient Greenland hunting traditions, whaling also functions as a cultural element. This is so not only because it is a cultural phenomenon in and of itself, but also because it embodies the 'first time' tradition, and because the distribution of the whale meat in accordance with the meat-gift tradition serves to give substance to a number of basic Greenlandic cultural values like generosity, benevolence, and interdependence. And last, but not least, the traditional communal eating of *muktuk*, be it immediately after landing the whale at the flensing ground or on festive occasions, has acquired an almost symbolical significance and has come to be regarded as something very special in the life of the Greenlanders.

Bibliography

Atusioq. 1985. Priser på grøndlandsk mad. I: *Atusioq* 5(5):11-13, November 1985. Atuakkiorfik.

Hansen, K.G. n.d. Newspaper quotation. Diary kept by K G Hansen. (Not published).

Hertz, O. 1977. *Ikerasarssuk — en boplads i Vestgrønland.* Copenhagen 1977, Nationalmuseet.

Petersen, R. 1985. The use of certain symbols in connection with Greenlandic identity. In *Native Power*, Brødsted et al. (eds.). Oslo, pp. 294-300.

Rink, H.J. 1982. *Eskimoiske Eventyr og Sagn.* (Reprint of original edition of 1866-1871). Volume 2, pp. 171-180, Copenhagen.

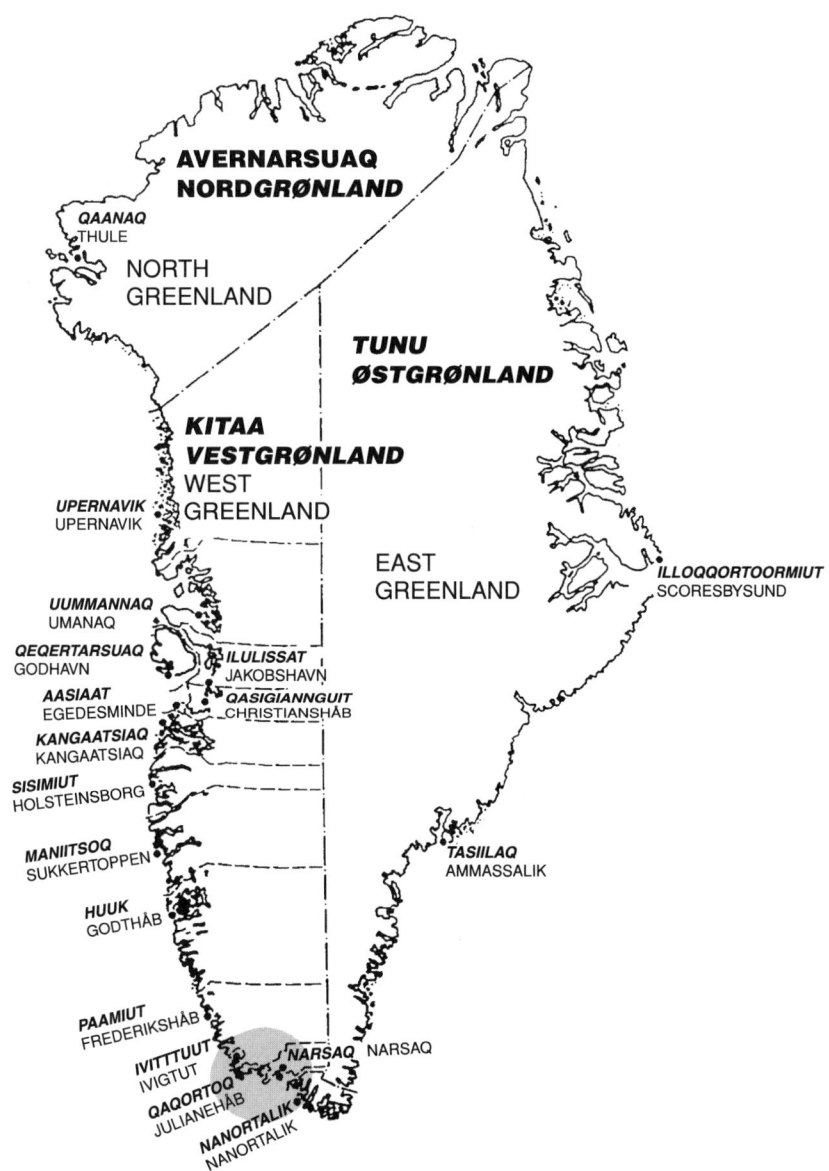

Figure 1. The Qaqortoq area in South Greenland.

Cutter Hunting of Minke Whale in Qaqortoq (Greenland)

Erling Josefsen
prepared for
the Greenland Home Rule Government
at the occasion of the 42nd Annual Meeting of the
International Whaling Commission
1990

Introduction

In Greenland, whaling takes place in many areas. Owing to geographical dispersion, local traditions, and present socioeconomic structures, whaling is quite different from one area to the next. For this reason, the importance of whaling for local communities is best described by means of case studies. Such studies illuminate the complex cultural context of which whaling in Greenland forms a part, and from which some general features can be deducted.[1]

Initially, case studies were planned in two locations in 1989-1990, Sarfaq and Qaqortoq. The locations selected represent a specific type of hunting practice (e.g., cutter hunting/hunting with small boats, fishing community/hunting community, town/settlement, etc.). This case study describes cutter hunting in the town of Qaqortoq (Figure 1.).

---

1. Editorial note: As background for this paper, the author provides an historical summary of whaling in Appendix I.

The municipality of Qaqortoq has 3,456 inhabitants of whom 2,892 were born in Greenland. The number of inhabitants in Greenland as a whole is 53,733. The town of Qaqortoq is considered a large town by Greenland standards. A look at the occupational profile of the municipality of Qaqortoq reveals that the number of persons who possess fishing and hunting licenses make up 23% of the inhabitants. Of these, 428 have green licenses (full-time) and 364 have red licenses (part-time). The national average is 29.1%.

Families included in this description of the cutter hunting of minke whales are called fishermen's families. The owner of the vessel is a full-time fisherman and, therefore, has a green hunting license. The members of the crew are also full-time fishermen.

Almost all the wives or co-habitants of these men have a job outside the home and, thus, contribute to the family's income. Without the incomes earned by the women, a family's income would not be sufficient for the family to survive over the long term. The owner of the vessel has a family to support, together with his co-habitant. They have three children; two attend school. With the exception of one, all the crew members also have families to support and two to three children each.

Whalers and Their Crew

There are two fishing vessels equipped with harpoon guns in Qaqortoq: the *Ani* (30 feet and 8.2 grt.) and the *Maren Olsen* (42 feet and 19.82 grt.), both built in 1965. The cutters are owned by two fishermen in Qaqortoq. Each cutter is equipped with a 50 mm harpoon gun. Detonating grenade harpoons are not yet used in Greenland. Only the so-called 'cold' harpoon, pending completion of the Greenland Home Rule's testing of the detonating grenade harpoons, is presently used. When the tests are completed, an adequate phasing-in period for these 'warm' harpoons will be required. Harpoon gunners strongly recommend the use of such harpoons, which they admit are very effective in securing and rapidly killing whales.

With these two fishing vessels the fishermen kill one minke whale each year out of a national quota of 60 whales, including the reserve quota of 10 minke whales in West Greenland. In addition, they catch one fin whale out of a national quota of 33. The national quota of 60 minke whales and 23 fin whales in 1989 has been fixed by the IWC.

The distribution of the municipality's quota is undertaken by the town council after consultation with the local association of fishermen and hunters, KNAPK. Thus, the town council, together with the KNAPK, distribute the quota to vessels with harpoon guns and small boats, respectively. However, the small-boat fishermen have to apply for dispensation first. The national quota for 1989 is distributed among 14 municipalities. Qaqortoq has been allocated eight minke whales which corresponds to 15% of the national quota. The town council of Qaqortoq, after consultation with the KNAPK, allocated one minke whale to the *Ani* out of the total municipal quota. The remaining seven minke whales are distributed among the hunters/fishermen who have applied for dispensation.

The *Ani* usually operates with a crew of four when fishing, whereas a crew of at least five is used for whale hunts. The other fishing vessel, the *Maren Olsen* usually has a crew of five when fishing and a crew of at least six persons when whale hunting. There is, however, no particular division of work or responsibility on board when the vessel is on a whale hunt. Besides the special task performed by the harpooner, jobs are allocated and performed almost as the situation demands. The master, who is usually also the owner of the vessel and its self-taught engineer, normally operates the harpoon gun, whereas the rest of the crew distribute and perform all the other jobs, e.g., mate, cook, coffee maker, and various odd jobs. It should be added that, by and large, most of the jobs are shared and carried out by all crew members.

Crew for the whale hunts are hired by the owner alone. The crew selected cannot be said to require or possess specific skills, and is not necessarily the permanent crew engaged in the daily fishing operations. The crew would sometimes consist of other fishermen and hunters, and perhaps also friends and acquaintances, not skilled in this profession, but nobody attaches much importance to this. Generally, no permanent crew is used for whale hunts every season. The selection of crew members does not follow any specific rules or requirements. However, on some occasions an experienced harpooner is sought, if, as in this case, no one has experience in performing this duty. One boat owner took over the position of harpooner in 1988 when he acquired the vessel. The selection of the crew is carried out in a relaxed atmosphere and everyone respects the choices made by the owner; they all know that it is not just a matter of participating, but of covering the local population's needs for whale meat. This method of crew selection underscores the impression that whaling cannot be called a commercial enterprise. Rather, in Qaqortoq it should be seen as a necessary and important activity, the purpose of which is first and foremost to meet the local population's needs for whale products.

The Whaling Season

Minke whales are found off the long coast of West Greenland, as far north as the Upernavik district. In late summer the minke is common as far north as Svartenhuk (72 N). In the southwest of Greenland, the minke is apparently most often found in spring and in autumn, which indicates that some whales migrate from the north to the south during the summer season. However, some minke whales seem to be stationary off the southwest coast during the entire summer season. Minke whales found in the spring are usually lean after a long winter in the open sea. (The fin whale usually arrives a little later than the minke whale.)

As mentioned, a number of minke whales can be found throughout the summer season, close to the coasts and especially in the fjords, which indicates that there is sufficient food for them in the area. From September to October minke whales are more often found offshore and their numbers increase considerably. In the summer season the minke whales appear individually. However, their behaviour changes dramatically in late autumn when they repair to coastal waters. Thereby, the fjord is almost emptied of whales as they tend to gather in herds offshore.

In the period from November to December the whales disappear from the coast to live in open sea, away from the fjords. Although whales can be found from May to October, whale hunting activities take place in the autumn months, late August to late October. There are several reasons why the activities are carried out in precisely this period.

Stable weather and good visibility are usually characteristic of the month of September in the south of Greenland, as opposed to the rest of the whaling season. These favourable conditions are among the most obvious and important reasons for the timing of the whale hunts. Successful whaling calls for good visibility and stable winds. Another important reason for the choice of this period is that, almost without exception, the whales are fatter and have more blubber in the late autumn months. The whales arrive in spring, when they are usually lean. During the spring and the summer they forage and put on fat in preparation for the forthcoming winter. The hunters and those in the population who eat whale meat are very much aware of this, as they all, without exception, prefer fat-rich whale meat to lean whale meat.

The fishing season is, of course, another decisive factor for the timing of the whale with hunts. In spring and summer the fishermen are primarily occupied with cod fishing, which takes place all along the west coast of Greenland. In autumn, beginning on the 1st of August, the fishing of salmon starts in the south of Greenland, and later in mid- and North Greenland. Salmon provides a good source of potential income throughout the autumn season. Therefore, all other activities are suspended and all efforts are devoted to the fishing of salmon. As salmon are expensive and subject to rigid quotas in the individual municipalities, the fishermen will invariably concentrate on fishing salmon until the quota has been reached. As it is also possible to sell to neighbouring municipalities, the salmon season can be prolonged for some fishermen, especially those employed on cutters such as the *Ani*.

Given the above conditions and circumstances, whaling activities take place in the period from mid-August to mid-October. The Greenland Home Rule order on minke whale hunting in 1989 stipulates that 'the hunting of minke whales within Greenland's fishery limits must exclusively take place in the period from 1 April to 31 December....'

The Hunt[2]

The hunting of whales has several stages. As a natural part of the hunt, the fishermen start by preparing for it in a number of ways. The harpoon gun and the harpoon are cleaned and prepared after not having been used for about a year. A spare harpoon is sent to the local shipyard to be straightened and welded, if necessary. The harpoon line is checked and, if warranted, replaced with a new line, whereupon it is carefully coiled to ensure that it does not get tangled up when the harpoon is fired. The deck is cleared of any remnants of fish, and tools such as fishing nets, pound nets, and fishing floats are removed and placed on the quay. In addition to the harpoon gun, other items are also brought onboard, e.g., a couple of 7.62 mm rifles, the only type authorized for whale hunting. But there are also other firearms onboard. Calibre 30-06, 222 rifles, saloon riles, and one or two shotguns. These weapons are exclusively used to shoot seals and birds. Furthermore, a large balloon float is placed on the deck together with a hand-held harpoon and is attached to the line.

---

2. Editor's note: The following sections contain a mix of narrative and description.

Erling Josefsen

The local population becomes more and more excited as the whaling season draws closer. Increasingly, local people contact the owner as well as the crew members to learn when they intend to go out whaling. Local fishermen who sail in the sea off Qaqortoq every day report more and more often that they have seen whales when out fishing. Such reports are very important for whale hunters. This kind of helpfulness and assistance provides the hunters with indispensable information which gives them the motivation required to go out and do their job as hunters. Without this local support they would obviously not be motivated to go whale hunting. This is an example of the cooperation that forms the basis of all whale hunting.

The *Ani* leaves Qaqortoq to go whale hunting at about 8 am. Over their radios other fishing vessels report that they have seen whales in the area, and at the same time, some small-boat fishermen report that they have seen the same whales in the same area, as well as other whales in other areas, depending on where they come from. The *Ani* sails to the closest area where whales have been reported. After about 30 minutes at sea, the first minke whale is sighted. Its behaviour shows that it is clearly looking for food. The hunt starts and the attention of the crew is focused on the pursuit of the whale. The first thing to do is to sail as close as possible to the whale and to place the vessel at a right angle to the whale and within range. At the same time, the maneuvering of the ship makes the minke whale grow accustomed to the noise of the engine. Everyone has to be patient and avoid reacting or doing anything in a hasty manner.

Unfortunately a minor problem has arisen. The vessel, and the minke whale, is in an area with a relatively high volume of boat traffic. Cutters, freighters, and some small-boat fishermen, who seem to be ignorant of the *Ani's* activities, are sailing about in the area. All this activity repeatedly disturbs the pursuit of the whale, which highly annoys the crew. As the traffic cannot easily be persuaded to move away from the area, and since it would be difficult to get within range, the crew decides to sail to another hunting ground in the hope that such disturbances will be avoided. Consequently, the *Ani* sails to another area in which there is a chance of finding other minke whales.

In the meantime, the crew have been successful in shooting a seal, which is cut up onboard the vessel. Some of the meat is cooked and eaten with relish for lunch. The rest of the seal is placed in a fish box, to be used as food during the rest of the whaling voyage.

A minke whale is discovered in the middle of the afternoon, but it seems to be extremely wary and later turns out to be a large, old whale. It is late in the afternoon and becoming dark. As no other whales have turned up in the area, the crew decides to go back and to return to the area the following day.

The Following Morning

Next morning at about 9 am the cutter sails to a coastal area northwest of Qaqortoq. At Kangerup Nuua, two fin whales and one humpback are observed. When we pass a stretch of open sea, we spot two other minke whales, but do not pursue them because of rough seas, probably created by the turn of the tide and windy weather further out on the open sea. Because of the unfavourable conditions the *Ani* sails to another area further south, between the islands not far from the island of Uummannaq.

On the way to the island of Uummannaq the crew shoot five black guillemots which are immediately plucked, cooked and served for lunch.

A minke whale is observed at noon and pursued. During the pursuit, three other minke whales are observed at close distance. Sometimes all four minke whales are close to one another, which actually makes the hunt more difficult. The behaviour of the minke whales clearly indicated that they were looking for food which could be seen from the fact that they frequently swam on their back and shoveled food into their mouth.

At about 6 pm the crew managed to sail close to a minke whale and to get within range for a strike. The harpoon is fired with an earsplitting roar. The whale has been hit, but the shot has not killed it. Two men carrying rifles jump into a small boat in order to do the final killing. Several shots are fired at the whale at very close range. Less than 10 minutes after the harpoon was first fired, the whale is dead. In these 10 minutes the activity onboard the vessel is hectic: attempts are made to haul in the whale while it is still alive. This is done to prevent the whale from sinking as deep as the length of the harpoon line. In the event that it does, it will require huge efforts to bring it to the surface. Unfortunately, the winch with which the whale could easily have been pulled up to the surface is out of order. Nevertheless, the crew manages to haul the whale to the vessel and secure it before it sinks. This was fairly easily done by sailing towards the whale.

At about 7 p.m. the vessel arrives at the flensing ground with the whale in tow. The flensing ground is located a little less than 8 km from Qaqortoq. In this area there is a small bay which seems to be ideal for flensing as the presence of a small, flat island seems a natural flensing place.

A strap is attached to the tail fin of the whale, and a small boat with an outboard motor slowly tugs the whale into shallow water. The strap with the line is attached to a rock overhang, and the whale is left in the shallow water, until the tide ebbs and it is free of the water and the flensing starts. Before the men leave the whale, they cut off part of the belly flesh. Thereby the offal is removed. This prevents the meat from being contaminated by the development of gases. The crew then leave the whale. They will come back when the water is low enough to start the flensing. Just before they leave they make a stew of the belly flesh and eat it with relish. The crew are happy after this successful day.

The Flensing

The flensing starts early the next morning at 4:30 a.m., 10 hours after the whale was caught. The whale is now out of the water. First, the *muktuk* is cut off in large slabs which are placed on the rock and cut into blocks, each weighing about 5-8 kg. These blocks are placed on the bare rock to cool. The blocks of meat are carefully placed side by side, never on top of each other. This accelerates the required cooling and prevents the meat from putrefying. It is well-known that whale meat spoils easily when it is warm and therefore should be treated with the utmost care. The harpoon is cut away and the meat around the harpoon, usually 40-50 kg is thrown away as it cannot be used for human consumption. Immediately before the flensing starts, the offal are removed and carried away from the flensing place.

At first, only four men take part in the flensing. Around 7 a.m. about half of the pieces cut off are transported to Qaqortoq in clean fish boxes placed on the deck of the cutter. At the quay in Qaqartoq the fish boxes with the whale products are unloaded and brought to the local open air market, which is popularly called 'the Board'. At the open air market the sale of other products has already started. The products sold here are various types of fish, lamb and heads of lamb, birds (kittiwake, black guillemot, etc.), and seal meat. The fish boxes with whale products are placed around the market place, whereupon the products are put on some tables already there. The sale starts at once. Two men cut and weigh the meat further and accept payment from the buyers.

Back at the flensing site four other men volunteer to do the rest of the flensing. Their offer is accepted without objection. For their voluntary participation in the flensing, the men are given remuneration in the form of whale products.

At about 10 a.m. the flensing resumes. The meat, the *muktuk* (with blubber), the tail fin, the dorsal fin, the belly flesh, and the heart are cut out in pieces of 5-10 kj each. These pieces are loaded on the cutter in fish boxes. The ribs are also cut out and placed on the deck. Remaining parts (i.e., the shreds of meat remaining on the spine) are left for anyone who passes buy and feels like doing a 'scraping off' of the whale. This meat is of prime quality and can be used for various stews. Normally scores of people will come from the town to look at the flensing and to secure a few kilograms of meat to bring back. But this flensing is unusual because of the weather and the time; the flensing started at 4:30 a.m. and at that time of the day most people are not yet out of bed. At the same time, the wind has increased in intensity which makes it difficult, if not impossible, to cross the strait to the flensing site. It could only be done by large vessels. Thus, no 'scrape-off' meat was collected on this occasion, leaving about 20-40 kg on the spine. The amount taken depends on how much energy people have left after flensing.

The entire flensing is done solely by means of large knives the length of a forearm. In general, no other tools are used. The flensing takes place in a relaxed atmosphere without any particular division of labour or supervision. Everyone, no matter whether they have participated in the hunt or have volunteered to help in the flensing take part willingly and energetically in the work. When the flensing is completed, the spine is split into smaller parts which are thrown into the sea together with the skull in order not to pollute the place. Thus prepared, the flensing ground can be used again and again without any prior removal of remnants from previous flensings.

Distribution of Whale Products

A substantial part of the whale products are distributed without any form of direct payment or remuneration. After the flensing some of the belly flesh, *muktuk*, heart, and tail and dorsal fins are distributed among those who participated in the catching of the whale and those who only participated in the flensing. Those directly involved in the flensing will give part of their shares to their families, friends, and/or acquaintances.

No participant will take more than what he needs for personal consumption and gifts. In total about 500-600 kg of whale products are distributed among the eight persons who participated in the flensing.

Every participant brings about 60-80 kg home, depending upon the size of their freezer and the amount of meat they intend to give away, whereas each voluntary participant is given 30-40 kg as remuneration for their work. Part of the share taken home for private consumption is usually given to family and friends as gifts.

During the sale in the open-air market, people sometimes come and politely ask for some meat. They are usually good friends or neighbours of those who participated in the hunt.

How many people directly benefit from the distribution of one whale is difficult to determine. A realistic estimate would be that about 30 households receive some form of remuneration for work carried out, either directly or in the form of gifts.

Sale of Whale Products

As soon as the whale products arrive at the open air market they go on sale. The buyers are exclusively local people who either arrive alone or together with their families. However, the local fish processing plant, Avataq, which is owned by the Greenland Home Rule, buys its share of the whale products, especially meat and *muktuk*, and belly flesh if there is any left.

There is no auction or any form of sales competition when the products are sold in the marketplace. The prices of the whale products have been fixed by the local association of fishermen and hunters in Qaqortoq, the KNAPK, at the beginning of the whaling season and prices are not changed during the season.

The price for pure whale meat is 30 kroner per kg, irrespective of the quality. The belly flesh is among the most expensive parts of the whale, costing 35 kroner a kg as it is considered to be a very fine delicacy. *Muktuk*, with or without a 5-7 cm layer of blubber costs the same, but no one really buys *muktuk* without the blubber which is often served as part of the fixings to, for instance, dried fish, dried seal or whale meat. Therefore, *muktuk* is almost always sold with the blubber. This time the tail, dorsal fins, and the heart were not sold at the market as they had already been distributed among the participants in the flensing.

The sale takes place in the open air market, with the exception of the sale to the Avataq plant on the quay. Avataq prices are slightly lower than market prices as it pays 10 kroner per kilogram for pure meat, irrespective of its quality. *Muktuk* and belly flesh are bought at 14 kroner and 10 kroner a kg, respectively. As mentioned, the buyers are exclusively local people. The average shopper usually buys 1-3 kilograms of mixed whale products. The belly flesh sells like hot cakes, and very often never reaches the vending table at the market; all of it has been ordered in advance. The local institutions (i.e., the hospital, the elders' residence and the Sulisartut Højskoliat, The Folk High School for Workers) buy on the same conditions as all other shoppers in the market. These institutions, however, buy a large part of the whale products at once, contrary to the ordinary consumer.

There are no rules applying to the distribution of the whale products between the ordinary consumer, various institutions, and the Avataq plant, but the general attitude among the hunters seems to be that the ordinary consumer shall have first priority so that they are sure to get their share. Therefore, orders are often given in the street, even before the whale hunts begin. In these times of restrictive whale quotas some try to secure their share of the whale products in advance. To date, plants such as the Avataq and the institutions have not requested beforehand a certain amount of whale products. The owner decides how many kilograms he wants to sell to them. Everybody realizes and respects that as many as possible get their share of the whale. Nobody gets more than they need.

Of the total catch of about 2.5 tons, two tons of whale products were sold; distributed as meat, belly flesh, and *muktuk*. About 40% of the catch was sold in the open air market, 24% to ordinary consumers and 16% to the institutions. The Avataq fish processing plant bought 40% and the balance (20%) went to private consumption. For the sake of good order, it should be noted that about 2.5% of the edible portion of the whale was wasted because of the harpoon and because the meat left on the spine could not be scraped off owing to some unfortunate circumstances. Benefits from the hunt, in regards to the distribution of meat after deduction of the shares sold to the institutions and the Avataq, reach as many as 300 households.

Erling Josefsen

Distribution and Consumption of Whale Products

At the market, fresh whale meat, belly flesh, and *muktuk* are sold as soon as a whale is landed. But such landings are becoming rare owing to the rigid quotas on whale hunting. This implies that whale products can no longer be taken into consideration as food reserves, despite their high nutritional value. Instead, much of the population has come to consider whale products as luxury goods where they used to be considered everyday food, and a necessary and nutritionally sound element in the ordinary diet. Whale products used to be on the menu at least once or twice a week. With the quantities available for each household today, whale products can only be served once a month. There is simply not enough whale meat for all, despite the huge demand among the local population.

The amount of whale meat secured by the individual household will either be cooked at once or deep frozen or dried in open air for future use. Before the meat is dried, it is cut into thin shreds or slices which are hung in the open air. The best time of the year for drying is the autumn and winter when the weather is colder and the flies have disappeared. Consequently, most people will freeze the meat and defrost it for the drying season. The meat is prepared either as a stew or fried in a frying pan. The belly flesh is usually cooked and served with potatoes. *Muktuk* is usually salted and eaten raw together with dried fish, dried seal meat or dried whale meat.

The Avataq fish processing plant is the only plant in Greenland that processes whale products for further distribution. Prime quality whale meat is cut into portions of steaks each weighing 400-1000 grams. Second-quality meat is prepared either for stews in portions of 500 grams, or is cut into thin shreds, machine-dried and packed in portions of 200 grams. The blubber from the *muktuk* is cut off, leaving the *muktuk* about 2 cm in thickness. It is then packed into portions of 700 grams. The belly flesh is also packed in portions of 500-700 grams each containing 3-4 pieces of belly flesh.

All pre-packed articles are checked and frozen, whereupon they are stored in the freezers of the plant. These frozen goods will later be distributed, primarily to local food stores (Greenland Trade, Brugsen Co-op, and private shops). A small percentage of these frozen goods could have been distributed to other towns on the coast, but the management of Avataq had no plans to do so as the local shops are able to buy all the products. The frozen whale products in the food shops are exclusively sold to local consumers.

Bibliography

Fly-Petersen, N. 1987. *Piniartorsuusimasup oqaluttuuppaanga*....
Gad, F. 1984. *Grønland*. Politikens Danmarkshistorie.
Lynge, H.A. 1982. *Piniarnilersaarutit.Meddelelser om Grønland* 1927-1966.

Erling Josefsen

Appendix I

Historical Background

Towards the end of the 18th century whaling for bowhead whales and humpbacks under the auspices of the Royal Greenland Trade was very common. The crews were European. The catches sometimes provided considerable amounts of blubber and baleen, especially at Holsteinsborg and Godhavn. During the war from 1807 to 1814, commercial whaling was suspended. Although it was later resumed, the catching of large whales had only minor importance as the 'stocks' had been thinned out as a result of ruthless exploitation by whalers from other countries. In the last half of the 19th century and the early 20th century no whaling took place under the auspices of the Royal Greenland Trade.

Greenlandic Natural Whaling of Bowhead and Humpback

The catching of bowhead and humpbacks by Greenlanders took place at irregular intervals and almost haphazardly. In order to catch the very large whales, many men were required. If a stranded whale was found, the flensing was a veritable orgy of blood, laughter, gluttony, fights, tragedy and comedy. If the whale had not stranded, but was asleep, *umiat* (women's boats) were used to approach it very quietly. The hunters then jumped 'onboard' the whale in their jump suits in order to thrust their harpoons into the blowholes of the whale. It was a highly dangerous sport. The *umiat* had to get away from the whale very quickly, and the hunters were thrown into the sea as a result of the furious reactions of the whale. That was why they wore jump suits, a kind of diver's dress, tied in a ring on the chest and preferably filled with air and coated with a thick layer of blubber. The dress fitted tight to the hunter and kept him warm and afloat. When the whale had been killed it was towed back, sometimes by a row of *umiat*. It was hauled into shallow water and then the flensing orgy started.

The techniques used in whale hunting were later developed into practices quite similar to those used by European whalers. In the Frederikshåb area a technique had been developed to catch humpbacks. It required that two spears and a harpoon with a line hit the whale simultaneously. A skin boat was created especially for use in whaling, but *umiat* were still used where no other boats were available. The whaleboat was wrapped in a very thin skin that made it light and fast. Two different types of oars were used during the hunt. Long and robust oars were used exclusively for the general propulsion, whereas the other type of oars was used when the boat came near to the whale. This type was usually shorter and less wide, in particular at the ends. Furthermore, the looms were padded to prevent noise which could attract the whale's attention.

Three men stood at the front of the skin boat, ready to lance and harpoon the whale. Their task was to attempt to hit the heart and the lungs. The target is between the fifth and the sixth rib, two handbreadths behind the undulated part of the front flipper. The humpback was approached as it lay floating with its front flippers deep in the water. The moment the spears hit the whale and the harpoon had been thrown it was very important to watch the line which was pulled by the whale at great speed. Sometimes the whale would die in a few minutes; other times it would dive or swim with the boat in tow for long distances. At times the whales' dives could be so powerful and deep that the line was not long enough. If so, it was sometimes necessary to cut the line with an ax that was always kept onboard to prevent mishap.

Before colonization, the humpback was caught by Greenlanders using *umiat*. When the Royal Greenland Trade had resumed the hunting of these whales with European crews, and later given it up again, it still continued in Frederikshåb where Greenlanders hunted humpbacks until 1923. In Godthab the hunting was resumed around the turn of the century, but it stopped at the same time as it did in Frederikshåb.

Whaling in the 20th Century

In 1924 the Danish Greenland Administration started modern whaling with the whale catcher S/S *Sonja* which hunted blue whales, fin whales, humpbacks and sperm whales. In addition to providing whale oil for the oil factory in Copenhagen, the purpose was to provide Greenlanders with meat for domestic consumption and 'leaf fat' for dog food in order to increase the dog population.

In the period from 1925 to 1928 the *Sonja* was accompanied by M/S *Sværdfisken* which was used as a flensing ship. From 1929 onwards the flensing was taken over by the Greenlanders. The whales caught were delivered to the colonies where the Greenlanders were given meat as payment for their flensing of the whale and packing of the blubber.

While whaling had been suspended during World War II the Danish Greenland Administration continued whaling with *Sonja* until the operation was eventually stopped in 1959. In 1951 the *Sonja* was replaced by another whale catcher, the S/S *Sonja Kaligtoq*.

The Tovqussaq Whaling Station

Headed by Mr. Hjalmar Andersen, a flensing station was created at the freezing plant at Tovqussaq in June 1954. The freezing plant and other plants in Greenland owned by Det Grønlandske Fiskerikompagni Ltd. were taken over by the Royal Greenland Trade.

The efficient flensing of the whales at Tovqussaq together with the instant freezing of the whale meat made whaling as a whole more profitable. As a result, whaling at Tovqussaq was considered economically viable, especially from a nutritional point of view. It thus became possible to distribute whale meat which was both rich in protein and other essential nutrients, and much in demand to the towns whose inhabitants needed this meat. The results of the pilot project in 1954 were so satisfactory that it was decided the operations would continue. It was considered extremely important to be able to contribute good, nutritious food in the form of whale meat and *muktuk* to the Greenland population.

Owing to a drop in whale oil prices, whaling in 1957-1958 resulted in rapidly growing deficits on the profit and loss accounts of both the whale catcher and the whaling station at Tovqussaq. In the spring of 1959, it was clear to the Royal Greenland Trade that the budget would be exceeded by a considerable degree because of the still unfavourable oil market and the apparent downward trend in whales landed. Even a doubling of the prices for frozen whale meat in Greenland, which was considered as an option, and a complete modernization of the whaling station at Tovqussaq would not be sufficient to balance the accounts. In consultation with the Board, the Royal Greenland Trade Department decided to cease its whaling activities in 1959. It was determined that whale meat for the Greenland population could be bought in Norway as part of the Department's ordinary supply service. The *S/S Sonja Kaligtuq* was sold, as no alternative ways of using it could be found.

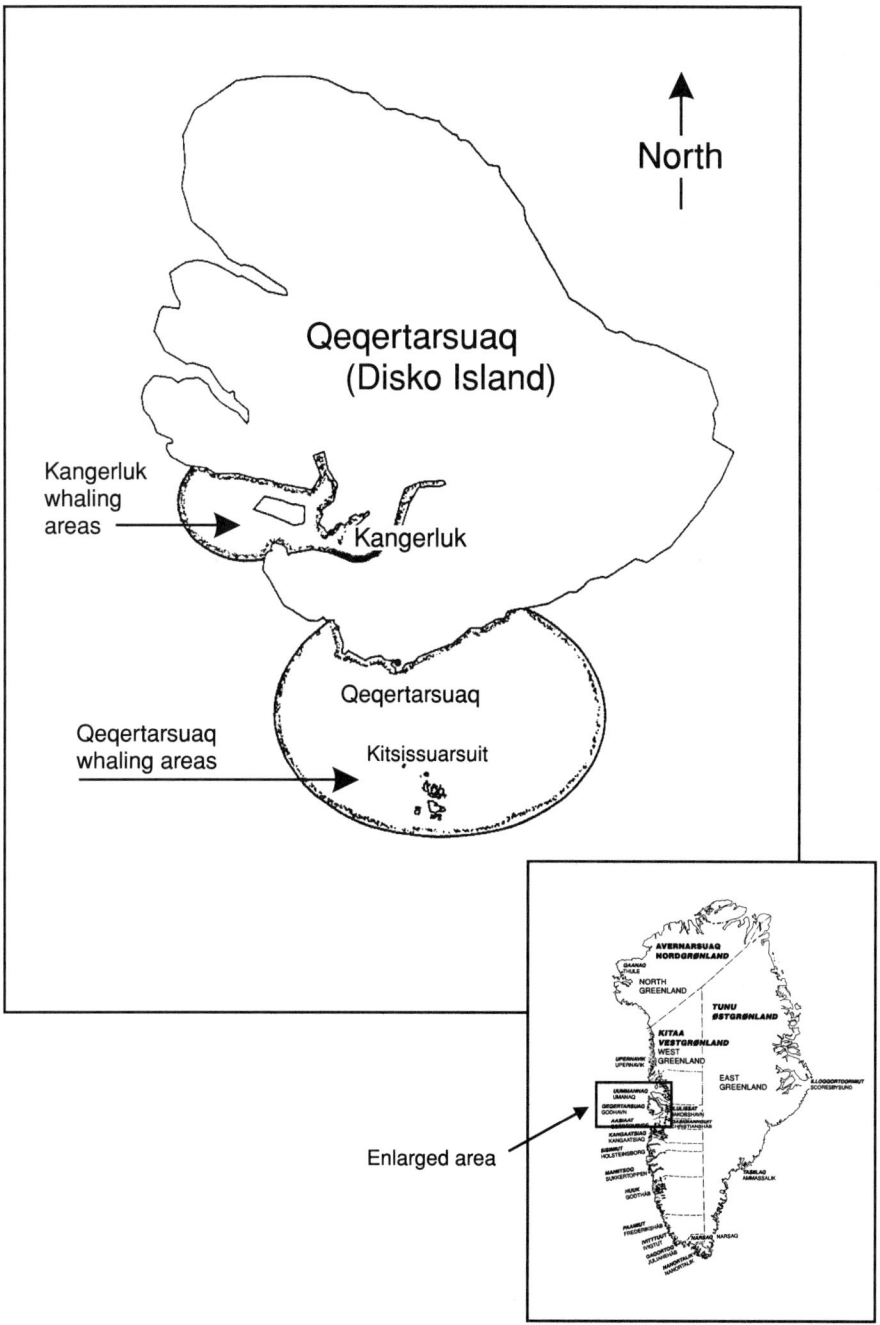

Figure 1. Qeqertarsuaq and the Disko Bay region.

Greenland Inuit Whaling in Qeqertarsuaq Kommune

Richard A. Caulfield
Department of Rural Development
University of Alaska, Fairbanks
1991

Nutritional and Sociocultural Significance of Whales in Qeqertarsuaq Kommune

Overview

Locally-obtained wild foods, referred to as *kalaalimerngit* in the West Greenlandic language, are of fundamental importance for Greenlanders in Qeqertarsuaq Kommune (Figure 1). Greenlanders differentiate these foods from Danish or other imported foods, which are called *qallunaamerngit* or 'white man's food.' *Kalaalimerngit* comprises a substantial part of household diets in Qeqertarsuaq and Kangerluk, and local residents actively participate in hunting and fishing in order to obtain these foods.

But the significance of Greenlandic foods extends far beyond the nutritional value that they offer. Greenlandic foods are intimately linked to Greenlandic identity. The processes of procuring, processing, preparing, and sharing Greenlandic foods help bind together nuclear and extended families, and indeed entire communities. The sharing of *kalaalimerngit* is a reflection of underlying systems of reciprocity and community solidarity which continue to be important in Inuit life today. For many Greenlanders, eating Greenlandic foods helps to reinforce a sense of cultural distinctiveness. As Larsen and Hansen (1990) point out:

> ...this distinction between "Danish food" and "Greenlandic food" is far more significant than a merely functional distinction referring to the origin of the food. Eating Greenlandic food is of great symbolic weight in determining whether a person is a true Greenlander...

Meat, *mattak*, and other foods from whales are very much a part of these cultural systems. In this section, I present data from Qeqertarsuaq Kommune about the significance of whale products in local diets, the sharing of Greenlandic foods within communities, and perspectives from local people about the continuing importance of *kalaalimerngit* in a changing world. Furthermore, I also discuss the complex and changing relationship between whaling and Greenlandic Inuit identity. In Qeqertarsuaq Kommune whaling is an integral part of a cultural dynamic centered around the harvest of renewable marine resources. Despite changes in technology and technique, whaling remains tied to Inuit cosmology and embodies a distinctive culturally-based relationship between humans and the natural world. It continues to be a vital part of oral traditions, and participation in whaling remains a source of prestige for hunters. In Qeqertarsuaq especially, symbols of whales and whaling's history are pervasive. Increasingly, whaling is contributing to an emerging sense of Greenlandic national identity.

Nutritional Significance of Minke and Fin Whale Products

Marine mammals, including whales, contribute to a high calorie diet which is highly desirable for hunting, fishing, and other outdoor activities in an Arctic climate. In Qeqertarsuaq Kommune, local residents rely upon a variety of wild and imported foods to maintain a balanced diet, but meat from marine mammals is considered the best source of nutrition. Seal meat, in particular, is viewed as 'real food.' Indeed, if one simply refers to 'meat' (*neqi*) in West Greenlandic, the assumption is that the reference is to seal meat.

The rich diversity of local foods available in Qeqertarsuaq and Kangerluk provides a sense of security for local people which is difficult for 'outsiders' to appreciate. While ships arrive all summer bringing imported foods which can be purchased at local stores, one can sense a special satisfaction among families as they share *kalaaliminertorneq*, a sort of smorgasbord of wild Greenlandic foods...

> ... Our host invited us to share *kalaalimerngit* with him and his family. We relaxed and drank tea while his young son went down to their outdoor cache to fetch the frozen foods. There was a flurry of activity in the kitchen, and shortly after we were invited to sit around a table literally covered with a rich variety of Greenlandic foods. There was *tuttu panertut* (dried caribou meat), *saarulliit panertut* (dried cod), *puisip tingua* (frozen seal liver), and *angmassat panertut* (dried whole capelin). We also had *tikaagulliip qiporaa* (belly flesh from minke whale), *quaq* (frozen seal meat), *tikaagulliip sarpia tarajugaq* (*mattak* from minke whale tail), *qilalukkap qaqortap mattaa* and *tinguanik imerlugu* (beluga *mattak* and frozen liver), and *tuttup iloqutai* (caribou mesentery fat).
>
> We tasted everything, tearing off pieces of dried meat or fish with our hands, or cutting off pieces of *mattak* with a sharp knife. We placed these pieces on a small wooden board in front of each of us, and ate small slivers of each food item as our hosts quietly did the same... (field notes, Qeqertarsuaq, 28 October 1989).

The importance of marine mammals in local diets and the diversity of wild foods eaten can be seen in Figure 2, which is based upon food consumption survey data from five households in Qeqertarsuaq. The largest category of wild food consumed was fish (23%), followed by seal and walrus meat (19%). Imported foods account for 16% of the meat and fish consumed, and minke whale meat and *mattak* comprised 6%. Of course, many factors (e.g., local environmental conditions, hunting quotas) can influence the types and amounts of meat or fish consumed, but the preponderance of locally-obtained marine resources in local diets is striking.

Seal meat is clearly the single most-preferred food (37%) among respondents to the Qeqertarsuaq Kommune Household Survey (Table 1). Interestingly, one-third of the households surveyed could not respond to a question about which Greenlandic foods were most preferred. These respondents simply said 'all of them', indicating that each of the foods has a place in seasonal diets. As one man said, 'during springtime I most prefer beluga and narwhal, but then in summer I like fish, seal meat, and minke whale. All of them fresh!'

Greenlanders believe that many Euroamericans either are not aware of or have a difficult time comprehending, the deeply-held feelings

Richard A. Caulfield

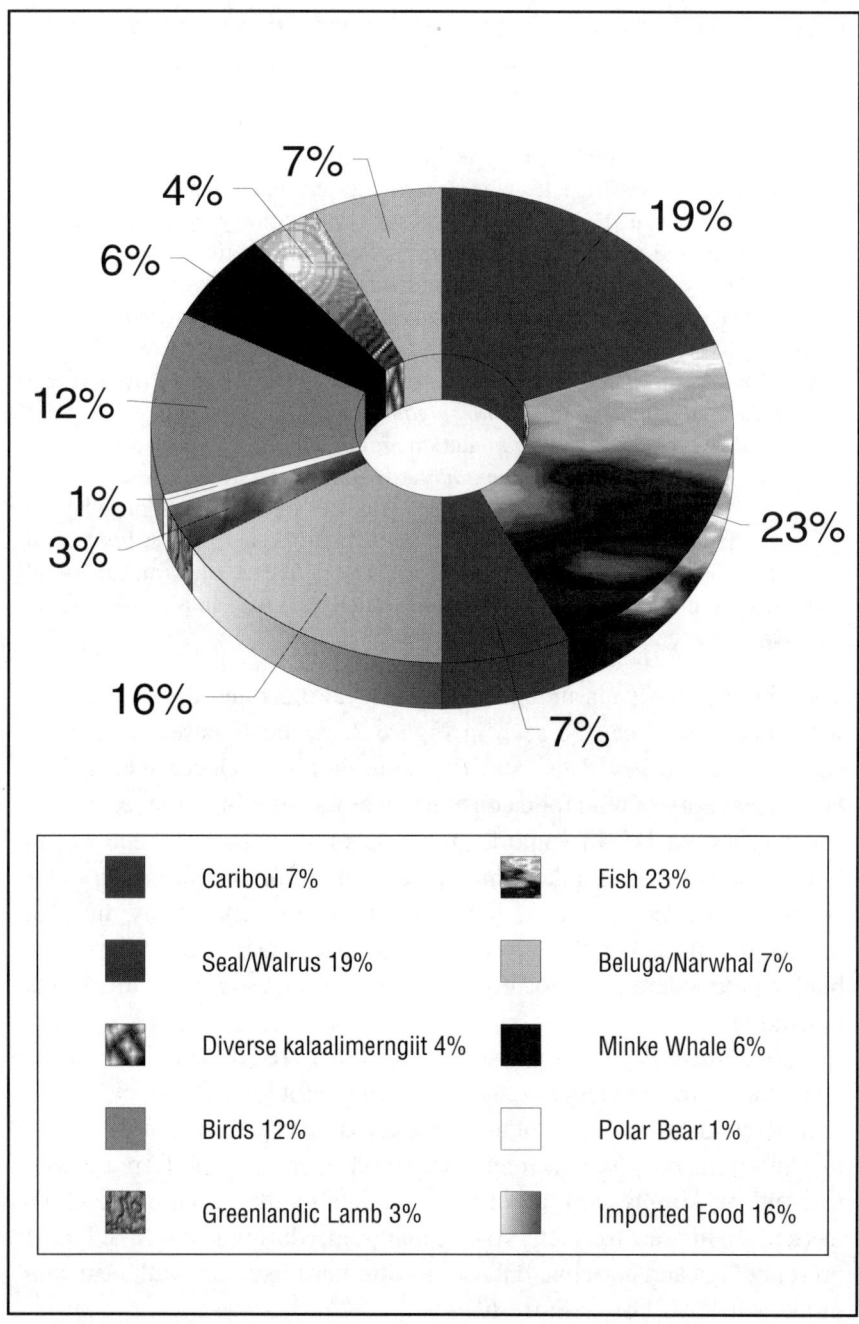

Figure 2. Principal meat and fish consumed by five selected households in Qeqertarsuaq Kommune, October 1989, January-April-July, 1990

Table 1.
Greenlandic Wild Foods Most Preferred by Household Members, Qeqertarsuaq Kommune Household Survey, 1989.

Food type	% of households
Seal	37
'all of them'	34
Minke whale	8
Caribou	7
Beluga	6
Fish	5
Birds	2

Greenlanders have about eating their own foods. As one young woman emphasized:

> We don't want to eat European food. When we eat European food, we don't feel full. Greenland is a cold place. When we eat European food, we get cold after one or two hours riding in a boat or traveling with a dog team. *Kalaalimerngit* is what we want to eat...especially for the old people and for the children, when they get sick. If they eat Greenlandic foods, it's better for them.

Sharing of Wild Foods in Households and Communities

Petersen (1988) describes the importance in Greenland over time of sharing wild foods. Until recently, sharing of foods on a generalized basis took place in smaller settlements, but people also participated in balanced exchanges where foods or commodities were bartered between households. These sharing and exchange networks fostered community solidarity and provided a sort of insurance against difficult times (ibid: 1-2). The meat-gift system helped support those who were not able to hunt for themselves, and reinforced a sense of reciprocity in communities.

However, these patterns of sharing began to change as settlements grew larger in the 20th century. As Kleivan (1984:609) points out, the growth of settlements meant that meat-gifts were increasingly restricted to relatives and close neighbors. These changes became particularly pronounced with the introduction of new and capital-intensive technology such as fishing vessels. Fishermen began to sell their catch in the settlement because of the considerable costs of the means of production.

Despite these changes, sharing of food remains an important part of everyday life in Qeqertarsuaq and Kangerluk. Data from the Qeqertarsuaq Kommune Household Survey indicate that 40% (n=25) of all households report that they always share wild foods with others. Ten percent (n=6) reported they often share, while 15% (n=9) said that they occasionally shared foods. Seven percent (n=4) reportedly shared only rarely, and 29% did not share at all.

For those households which do share, virtually all, 98% (n=44), share with immediate family in their own community, while 78% (n=35) share with extended family in their community. Only 40% (n=18) share with friends or acquaintances in their own community.

Households are much less inclined to share with people outside of their community owing to logistical considerations, although 56% (n=25) did so with immediate family members in other communities. Twenty-four percent (n=11) shared with extended family members in other communities, and only 11% (n=5) shared with friends and acquaintances in other communities.

In Qeqertarsuaq and Kangerluk, gifts of meat or fish (called *pagugut*) are an important part of community life and are commonly shared. Household members might simply share foods with those whom they like or who have helped them in some way (*qujagisaqarneq*). Naming relationships (*atsiaqarneq*) are common in Greenland (Langgaard 1986:304), and wild foods are frequently shared with the person with whom one has a name relationship. The named person, or *atsiaq*, will commonly receive meat-gifts from the family of the person for which he is named. Furthermore, the *atsiaq* will refer to the father in the other household as his *ataatakulooq*, and his wife as *anaanakulooq* (both terms of endearment appended to father and mother, respectively) and will commonly spend time with that family.

Meat-gifts are also shared with *ilaquttat*, or close family members such as parents, grandparents, or cousins, in other households. They might send gifts to siblings who are living away from home, in Nuuk or in Denmark for example. Or, hunters might share meat-gifts with those unable to hunt for themselves (*pilersuisoqanngitsut*). For example, in Qeqertarsuaq, hunters frequently share meat, fish, and *mattak* with elders who have no children to hunt for them. In some cases, meat gifts are also made to *piniuteqanngitsut,* those lacking the means to hunt or fish for themselves. For example, active hunters in Qeqertarsuaq gave meat to one household because they didn't have a functional outboard motor for a period of time. The hunters realized the difficulties that this created for procuring wild foods, and brought by gifts of meat until a new motor was available.

In Qeqertarsuaq and Kangerluk, the types of meat-gifts differ depending upon the sex of the recipient. For example, boys usually receive the *puisip tajarneq,* or the front flipper and claws of a seal. This gift is said to give the boy strength in hunting. In contrast, girls receive the *puisip pamialluk,* or the lower vertebrae and coccyx (tail bone) of a seal. Gifts such as these are made without consideration of how much money or material wealth a household possesses. The owner of a shrimp trawler, a full-time employee of KNI, and a self-employed fisherman can all be equal participants in this sharing network.

However, the system of sharing wild foods is clearly changing. Seventy-six percent (n=47) of households interviewed noted that there is less sharing now than 20 years ago. Thirteen percent (n=8) said that sharing now was about the same, while 11 (n=7) had no response. The major reason cited for this change is that more people sell hunting and fishing products today than they did 20 years ago, because they need the money. Most people said that the number of households sharing hasn't changed as much as the extent of the sharing. Thus, households are more likely now to limit their sharing to their immediate or close extended family members.

Wild Foods in Household and Community Celebrations

Whaling festivals of the type held by Inuit hunters in Alaska do not exist in Qeqertarsuaq Kommune or elsewhere in Greenland. However, given the central role which Greenlandic foods have in sharing networks, it should be no surprise that whale meat, *mattak,* and *qiporaq* are integral parts of household and community celebrations. The types of celebrations which take place range from an extended family simply getting together to have a picnic outdoors on a warm sunny summer day, to a community-wide fest, where Greenlandic foods are shared among all participants. Greenlanders in Qeqertarsuaq love to share wild foods outdoors with their families on summer days. Entire extended families, from elders to babies, gather in sheltered locations on the outskirts of town to cook over a small fire made of dwarf birch and crowberry. Whale meat, fried in butter, is a favorite food on these occasions.

Few household celebrations are as distinctive as the Greenlandic *kaffemik*. These gatherings are held throughout the year to celebrate events like birthdays, anniversaries, baptisms, or confirmations. On some days, there may be as many as four or five *kaffemiks* in town on a single afternoon. At a *kaffemik*, guests simply stop by at any time during the afternoon or early evening. After coming into the house and giving special greetings to the honored person and his or her family, guests sit at a table brightly decorated with candles, and brimming with cakes and sweets. Many of the foods served are of Danish origin, including layer cakes and cookies. But depending upon who the guests are, local foods such as beluga *mattak* and dried fish are also served. Guests are welcome to take small pieces of caribou fat from a china bowl and place it into their tea, making a delicious broth. The contrast of Danish sweets and *kalaalimerngit* may appear odd, but somehow the blending of the two isn't as strange as it seems. Field notes from January 1990 describe this juxtaposition of Inuit and European foods.

> We went to two *kaffemiks* today, one to celebrate a couple's 25th anniversary, and the other for the 75th birthday of a respected hunter. After knocking and then entering the home of the hunter, we squeezed into the tiny entry cluttered with hanging parkas and scattered *mukluks* from guests who'd already arrived. We slipped off our heavy boots and clothes and shook hands with the hunter and his family, saying *pilluarit* (congratulations) to each. We were invited to sit down at the table decorated with candles and a white tablecloth, while the hunter's wife served us Danish buttered bolle (rolls) and coffee. We then helped ourselves to some delicious layer cakes and cookies.
>
> As we ate, the hunter's son came noisily into the kitchen hauling a seal which he had just caught. Guests admired the seal and commented on the ability of the hunter who had captured it. After finishing our cakes, we were welcomed back into the kitchen where we were invited to take turns carving off *mattak* from the flipper of a narwhal. Sharing a single sharp knife, we picked away at the flipper, which was perched on the kitchen counter next to the coffee percolator (field notes, Qeqertarsuaq, 18 January 1990).

Special meals of *kalaalimerngit* are also served when special guests arrive from other communities. Local people believe that a proper welcome for visiting family members or Greenlandic dignitaries includes a generous serving of wild foods. Community-wide gatherings take several forms. Frequently, community dances (fests), are held in the *katersortarfilc*, or community hall, to commemorate special events or to raise money for a worthy cause. The dances usually begin about 9:00 p.m. and last into the early morning hours. Quite often, participants share Greenlandic foods such as dried whale meat, *mattak*, and dried capelin just after midnight as the evening winds down.

Greenland's national day, June 21, is also a day of community-wide celebrating. In Qeqertarsuaq, the entire community gathers in the afternoon and evening on the outskirts of town near the Arctic Station to listen to Greenlandic music, play games, and share wild foods, including whale meat, cooked over an outdoor fire.

Whaling and Greenlandic Culture Identity

Greenlandic society today continues to undergo substantial changes as it becomes more and more integrated with the world economy. World prices for shrimp and other marine products, seal skins, oil, and minerals influence the lives of all Greenlanders, even in the smallest settlements. While these changes ripple through contemporary Greenlandic society, other factors provide Greenlanders with a sense of continuity with their past, and a sense of what it is to be a Greenlander in a changing world. Chief among these is the 'harvest' of marine mammals, including whales.

Participation in marine mammal hunting provides Greenlanders with a valued connection to a hunting way of life reaching back over 4,000 years. The Inuit of Greenland share this connection with fellow Inuit in Canada, Alaska, and the Soviet Far East. As a people, the Inuit are 'by tradition the most hunting-oriented of all human groups, because their environment provides very few non-animal resources' (Donovan 1982:40). Therefore, whaling and the sharing of food from whales are manifestations of cultural continuity with Inuit traditions of great time depth (Petersen 1987, 1988).

As Dahl (1987) points out, whaling and other marine mammal hunting in Greenland today serves complex integrative and cultural functions. In Qeqertarsuaq Kommune the social organization of whaling remains closely tied to kinship relationships between hunters. When extended family members participate in catching a whale, whether on a fishing vessel or in a collective hunt, bonds of kinship are strengthened through shared experience. This participation initiates a host of social and cultural interactions. As family members assist in flensing, meat and *mattak* are shared with others, and stories are told about the hunt. The language of these interactions between kin is *Kalaallisut*. Much has been written about the links between language and culture in Greenland focusing upon the remarkable vibrancy of the West Greenlandic language and its central role in contemporary Greenlandic identity. When hunters talk excitedly among themselves over walkie-talkies about the sighting of a whale, they share that discovery with a discreet cultural group because of shared language. Whaling and hunting generally reinforce these ties to language and employ words and phrases about the hunt which developed far back in Inuit history.

Similarly, whaling requires sharing of knowledge, values, and beliefs. When an elder hunter and experienced whaler participates, his knowledge about whale behavior and characteristics is respected. As is typical in Inuit society, participants defer to the elder because of the knowledge and experience he brings to the hunt. Culturally-appropriate behaviour in whaling is transmitted to others through example. Admonitions against waste and an emphasis on thoroughly using an animal are stressed as older and younger hunters work alongside each other when hunting and processing animals. In this manner, Inuit knowledge and belief systems are passed on to younger hunters through example and experience.

Participation in whaling is a source of prestige which validates the continuing importance of hunting in Greenlandic society. Greenlandic men speak with barely-muted pride when they describe their participation in whaling. This pride is a distinctly male phenomenon. Through the family's celebration of a boy's first seal, to the excitement of a community upon hearing that a whale has been taken, the role of the hunter, who is nearly always male, continues to be validated in Greenlandic communities. This validation may be especially important in communities, like Qeqertarsuaq, where many men engage in wage employment. Participation in a collective hunt provides a touchstone validating one's capabilities as a hunter, even as one spends much of the day in an office or a shrimp processing plant.

At the community level, whaling and marine mammal hunting generally provide economic insurance and a sense of social security. Men who are seasonally unemployed, or who can't find work at all, know that they and their family can survive because they can hunt seals and whales for food and for small amounts of cash for essentials. Greenlandic foods are valued highly, and can provide essential elements to household diets which could not be supplemented with store-bought foods.

The symbols of whales and whaling, evident throughout Qeqertarsuaq Kommune, demonstrate the vital link between whaling and community solidarity. For example, some of the most visible historic structures in Qeqertarsuaq are those associated with the community's early years as a whaling center. This includes the beehive-shaped lookout on Qaqqaliaq, the small promontory south of Qeqertarsuaq which overlooks the sea. The structure originally served as protected lookout for hunters watching for whales, and was built with four Greenland right whale jawbones. It blew down in a powerful storm, and efforts were made to rebuild it in time for the kommune's 200 year jubilee in 1973. In 1968, local authorities obtained dispensation to catch two Greenland right whales so that this historic structure could be replaced. However, only one whale was caught, and the reconstructed lookout had to be built of wood.

The jawbones from the one whale caught now serve as a ceremonial archway at Kongebroen, the dock for visiting royalty and dignitaries in Qeqertarsuaq's harbor. This structure is a very visible reminder of the community's long history in whaling. Furthermore, the ceremonial shield for Qeqertarsuaq Kommune itself, used widely to identify kommune buildings, vehicles, and correspondence, is that of a Greenland right whale.

Whales and whaling have for many years provided a sense of regional identity for Qeqertarsuaq and for other Disko Bay communities. Each of Greenland's major regions is distinctive, owing largely to distance, economic status, local language dialects, and history. Disko Bay is known for its abundance of whales, and as a place where whaling has been important for generations.

Increasingly, whaling is contributing to an emerging sense of Greenlandic national identity. As conflict between Greenland and other nations continues over marine mammal use generally, and whaling in particular, hunters in Qeqertarsuaq Kommune and throughout Greenland feel a growing sense of solidarity in fighting to protect what they view as a fundamental right to 'harvest' marine mammals for their livelihood. This common concern also contributes to a pan-Inuit solidarity, with hunters in Greenland joining those in Alaska at meetings of the IWC to emphasize the strong links between whaling and Inuit culture.

Richard A. Caulfield

Analysis of Contemporary Whaling Issues in Qeqertarsuaq Kommune

Overview

In recent years, critics of Greenlandic whaling have asserted that profit maximization, commoditization, and capital intensification may be taking place. At the same time, Greenlanders have voiced dissatisfaction with reductions in IWC quotas, which have sharply reduced minke whale allocations for local hunters and forced the Home Rule government to implement regulations limiting entry to whaling. Local hunters say that their needs for whale meat, *mattak*, and *qiporaq* are not being met, and that larger quotas are needed. In this section, I analyze these issues using data from Qeqertarsuaq Kommune, and evaluate the impact of whaling regulatory regimes external to the kommune (e.g., the IWC and Home Rule government) on local communities and whaling practices. In particular, I focus upon the impact which recent regulations limiting entry to whaling has on social differentiation within local communities.

Profit Maximization and Commodification in Whaling

Data from this study demonstrate that neither fishing vessel whaling nor collective whaling in Qeqertarsuaq or Kangerluk are carried out on a profit maximizing basis. The two fishing vessels actively involved in whaling are used principally for shrimp trawling. Whaling is an economically marginal activity for both vessels. According to vessel owners, gross income from whaling in recent years has amounted to no more than about 10% of the total gross income. As an illustration, an older-style cutter caught a single minke whale in 1988. In that year, the vessel's gross income from shrimp trawling was about $50,000 (270,000 Dkk). The vessel was involved in whaling for less than one week. Sales of meat and *mattak* from the single whale amounted to about $4,000 (25,000 Dkk), providing a gross income of about $54,000 from all operations. Thus, sales of whale products amounted to about 8% of gross income for 1988.

Similarly, a newer-style fishing vessel grossed about $550,000 (3 million Dkk) from all operations in 1989, according to its owners. Proceeds from the sale of whale meat, *mattak* and *qiporaq* were about 1% of gross income. Sale of about 1300 kg of minke whale products at the local *kalaaliaraq* (market) brought in just over $5,500. Expenses directly associated with the hunt (diesel fuel, powder and caps for the harpoon cannon, food, etc.) were estimated to be about $900 (5000 Dkk). Therefore, the net return from whaling was approximately $4,600.

Cash plays an even smaller role in collective whaling. Customary distribution practices often leave each hunter with relatively small amounts of meat and *mattak*. For example, if we use survey data averages for the numbers of participants and skiffs in recent collective hunts, 2000 kg of meat and *mattak* would be divided between 16 skiffs and 30 hunters. Each skiff would receive about 125 kg of meat and *mattak*, which would be evenly divided between two hunters, equaling just over 60 kg each. If three-fourths of this were meat and one-fourth *mattak* the individual hunter would receive the equivalent of about $340 worth of whale products. However, the costs of participating in collective whaling are relatively small. The average hunter spends just over $50.00 for fuel and ammunition in this type of hunt. Thus, if one considers only the direct costs of participation in collective whaling (not including basic costs of equipment or time away from other work), collective whaling does bring positive economic benefits.

In the 1960's and 1970's, some Qeqertarsuaq hunters commonly sold whale meat and *mattak* through KTU's predecessor (KGH) and through private sales. Yet, even during that period, cultural factors restrained any significant move toward profit maximization. As one long time observer in Qeqertarsuaq pointed out, whaling in the 1960's and 1970's never became highly efficient. There was so much prestige associated with whaling that Greenlanders were reluctant to treat whales as a simple commodity and to turn a 'harvested' whale over to others for processing. Had the local hunters divided up the tasks of catching and processing, according to this analysis, the resulting specialization would have made a much more efficient operation. This was not done. Those catching the whales were determined to take the time to be personally involved in all phases of catching and processing, thus foregoing any chance of maximizing production.

While an avenue exists for selling minke and fin whale products as a commodity through KTU in Qeqertarsuaq for resale in other areas of Greenland, it has had little use in recent years (Bjerregard 1990). Sales of minke and fin whale products have been limited to the local *kalaaliaraq* and occasionally to private sales. Data from the period of fieldwork in 1989 and 1990, and from interviews with hunters and others, suggest that these sales have been relatively small.

Capital Intensification and New Technology in Whaling

Fishing vessels involved in whaling, like the rest of Qeqertarsuaq's fishing fleet, have been made more efficient and powerful in recent years to meet the demands of the commercial shrimp and fishing industries (Berthelsen *et al.* 1989:15). However, the present overabundance of large shrimp trawlers in Greenland makes it unlikely that new vessels of that type will be built in the near future. Older fishing vessels, designed for multiple species harvests of shrimp, cod, and other fish species, are but slowly being replaced with more efficient vessels of about the same size.

However, despite these changes, the technology used for whaling, principally the harpoon cannon mounted on the bow of a fishing vessel, is virtually the same as that used 40 years ago. There has been virtually no capital intensification directly associated with whaling in Qeqortarsuaq's fishing fleet.

Interestingly, new investments in whaling technology are likely to come not for economic reasons but because of international concerns about 'humane' killing of whales. IWC concerns about improving the efficiency of minke and fin whale hunts by fishing vessels have led the Greenland Home Rule government, KNAPK, and selected fishing vessel owners to test the Norwegian-designed penthrite grenade, the so-called 'hot' harpoon which kills a whale within seconds of a strike. Vessel owners in Qeqertarsuaq seem favourably inclined toward the use of this technology, which is estimated to cost several hundred dollars per grenade. Yet the use of this technology requires changes in whaling practices and involves further costs and training. Further specialization will be required so that vessel owners and crews can safely and efficiently use the exploding grenade. Some Greenlanders believe that it is ironic that international pressures for humane killing of whales lead to greater specialization and expanded use of technology in Inuit whaling. Nevertheless, concerted efforts are being made to introduce this new technology as quickly as possible.

Impact of External Regulatory Regimes on Whaling in Qeqertarsuaq Kommune

Recent IWC actions are having significant effects upon the social and cultural foundations of whaling in Qeqertarsuaq Kommune. IWC quotas and the regulations implementing them have reduced significantly the number of whales caught, placed a strain on culturally-based hunting practices, and fostered increased dissatisfaction with whaling management regimes.

IWC minke and fin whale quotas for Greenland are allocated annually among kommunes in West and East Greenland by the Home Rule government in consultation with KNAPK and KANUKOKA, the national kommune organization. Criteria used in the most recent allocation included kommune population size, availability of suitable vessels with harpoon cannons, and the availability of economic alternatives. Quotas have been allocated to the kommunes, and the division of these allocations between vessels and collective hunters has been placed in the hands of the kommune council.

In addition, the Landsting (Greenland Parliament) has enacted laws governing whaling within Greenland. The Home Rule government implements these laws through regulation. These laws and regulations include requirements for reporting whale catches, requirements for the types of equipment and procedures which must be used in whaling, and penalties for violating applicable statutes. For example, Greenland law requires that at least two fishing vessels share in the taking of a fin whale to minimize any chance of loss and to maximize the local benefit from low IWC quotas. Fishing vessels involved in whaling are required to use a harpoon cannon of 50 mm or larger.

Special dispensation must be received from the Home Rule government before a collective hunt for minke whales can take place. Once dispensation is received, a minimum of five skiffs must participate, each of which must have a harpoon on board to minimize chances of losing whales. Only rifles of .30-caliber or greater are to be used in collective whaling.

Whaling is limited only to those hunters who hold a full-time hunting and fishing license, who reside in Greenland, and who have close affiliation with Greenlandic society. In a recent development initiated in 1990, the Home Rule government has begun strict supervision of whale allocations and catches through a system of whaling licenses. Hunters who are allocated minke or fin whale must report catching a whale to kommune authorities and obtain a stamp on their license before any whale products can be sold.

Qeqertarsuaq Kommune received a quota of six minke whales and no fin whales in 1990. In 1989 it received a quota of two minke whales and one fin whale, while in 1988 it received a quota of five minkes and shared, along with six other kommunes, in a general quota of five fin whales. The IWC quota for minke whales in 1989 was the lowest it had ever been (60 for all of West Greenland). Qeqertarsuaq's quota of two minkes was divided between the Qeqertarsuaq and Kangerluk.

Residents of Qeqertarsuaq and Kangerluk believe that quota allocations for minke and fin whale in recent years have been far too low to meet local requirements. Eighty-six percent of all households surveyed (n=62) could not obtain enough minke whale products in 1989, while 76% could not get enough fin whale. Households which usually participate in whaling cited IWC quotas as the principle factor limiting their ability to whale in 1989. Fifty-five percent said that IWC quotas were responsible. Eighteen percent reported that they were too busy with wage employment, and 11% reported that they were too busy with other hunting and fishing, principally commercial fishing.

Until the advent of collective whaling in the 1970's, Greenlandic whaling was limited to those who owned fishing vessels that could be outfitted with harpoon cannons. All others who desired whale products generally had to receive them as gifts from vessel owners or to purchase the products. Collective whaling enabled those without a whaling vessel to catch minke whales. The technology involved was limited to that owned by many households, and the costs were relatively low. Thus, the emergence of collective whaling helped ease any tensions between those who owned more expensive and sophisticated means of production and those who did not.

The imposition of reduced IWC quotas and the stricter monitoring of whale catches by the Home Rule government in 1990 has helped to raise concerns among some hunters about access to whaling. At one time, collective whaling was open to nearly any hunter who had the necessary equipment. Even those with wage employment could drop whatever they were doing and jump in a skiff when word came over the radio that a minke whale had been sighted. The shared experiences of hunting a whale and the opportunity to distribute meat and *mattak* often overshadowed the demands of a work schedule. However, current quotas and regulations exclude many part-time hunters who have a long personal history of involvement in whaling, a desire to procure whale products for their households, and a culturally-based belief that whaling should be accessible to all Greenlanders.

These concerns are compounded by the fact that many of those with full-time hunting and fishing licenses in Qeqertarsuaq Kommune have them because they are employed on commercial shrimp trawlers. Thus, a deckhand on a trawler can be eligible to catch whales, but a part-time painter, whose small income is supplemented by hunting and fishing, is ineligible.

These tensions and perceived inequities contribute to a growing sense of frustration among, and alienation of, hunters in Qeqertarsuaq and Kangerluk about whaling management regimes. Hunters attribute these problems largely to IWC quota restrictions, which they feel are based upon factors other than simply biological necessity. Recent infractions of whaling regulations in other West Greenlandic kommunes could well be a reflection of this alienation (Anon. 1990).

Conclusion

Contemporary Greenlandic Inuit whaling for minke and fin whales in Qeqertarsuaq Kommune provides important economic, sociocultural, and nutritional benefits to local communities. Contemporary whaling is part of Inuit socioeconomic and cultural systems which for generations have survived by harvesting and utilizing a wide variety of marine resources. While ecological changes, colonial policies, and expanding interaction with the world economy have brought many changes, Greenlanders have demonstrated an ability to accommodate new forms of technology and new harvest practices while retaining a sense of continuity with their past. As Dahl (1987) points out, this ability to mobilize traditional values in meeting the demands of modern times may be a principal defining characteristic of Greenlandic Inuit society.

The technology, processes, and even the whale species sought in Greenlandic whaling have changed substantially from earlier times. Cash now plays a role in the contemporary whaling complex, both as a means of purchasing necessary equipment and as a means for keeping distribution channels for hunting products open. Furthermore, avenues exist for hunters to sell some or all of their whale meat, *mattak*, and other products for distribution to other Inuit communities in Greenland.

However, the data from Qeqertarsuaq Kommune demonstrate that cash has a limited role, and that relatively few whale products enter into wider markets beyond the communities themselves. What is most significant is the continuing importance of whaling within a multi-species 'harvest' pattern. Participation in whaling, particularly for minke whales, is widespread, and whale products are integral parts of local diets and of sharing networks. Whale products are commonly used in family and community celebrations, and whaling continues to be important to individual, community, and national identity. Catching a whale brings prestige to the hunter, who for many continues to epitomize true Greenlandic values and traditions.

Recent whaling quotas have raised concerns about the socioeconomic and cultural foundations of whaling in Qeqertarsuaq Kommune. Quota reductions and stricter regulations limited access to whaling in order to protect those hunters most dependent upon hunting and fishing for their livelihood. In Qeqertarsuaq, this policy has the effect of disenfranchising many from whaling, and contributing to a process of social differentiation.

Greenlandic hunters share the concern of many about the continued viability of North Atlantic whale 'stocks.' They also believe that harvests of marine mammals are vital if the goal of sustainable development is to be achieved (World Commission on Environment and Development 1987, Griffiths and Young 1989). Hunters fear that distant political forces will curtail aboriginal whaling, in violation of what they view as fundamental cultural and human rights. While they recognize that politics inevitably plays a role in the management of whaling, they share Freeman's hope (1990:115) that Inuit cultural perspectives about whaling will be respected.

> The hope is that more balance, in the form of greater tolerance, respect and appreciation of others' cultural perspectives and needs will characterize future discussions. As important as biodiversity is to the maintenance of ecological processes in the biosphere, so is cultural diversity important to the ongoing development and vitality of the biosphere and with it the longterm survival of humanity.

Bibliography

Anon. 1990. Korrekte oplysninger om hvalfangst. *Sermitsiak'*, nr. 3, 19 January 1990, p. 26.

Bjerregård, N. 1990. Personal communication. Qeqertarsuaq, 15 January 1990.

Dahl, J. 1987. Tradition og kultur i den grønlandske naturudnyttelse (Tradition and culture in Greenlanders' utilization of nature). *Grønland* 35(10):295-304.

Donovan, G. (ed.). 1982. *Aboriginal/subsistence whaling*. International Whaling Commission. Special Issue 4. Cambridge.

Freeman, M.M.R. 1990. A commentary on political issues with regard to contemporary whaling. *North Atlantic Studies* 2 (1-2): 106-116.

Griffiths, F. and O.R. Young. 1989. 'Sustainable development and the Arctic.' Paper prepared by co-chairs of the Working Group on Arctic International

Relations, based upon a session held in Ilulissat and Nuuk, Greenland, 20-24 April 1989.

Kleivan, I. 1984. West Greenland before 1950. *Handbook of North American Indians, Arctic*. Washington, DC: Smithsonian, pp. 595-621.

Langgaard, P. 1986. Modernization and traditional interpersonal relations in a small Greenlandic community: a case study from southern Greenland. *Arctic Anthropology* 23 (1-2): 299-314.

Larsen, S.E. and K.G. Hansen. 1990. *Inuit and whales at Sarfaq (Greenland)*. Paper submitted to the International Whaling Commission. (TC/42/SEST 4).

Petersen, R. 1988. 'On traditional and present distribution channels in the subsistence hunting in Greenland.' Paper presented at the 1988 Inuit Studies Conference, Copenhagen.

World Commission on Environment and Development. 1987. *Our Common Future*. Oxford University Press, Oxford.

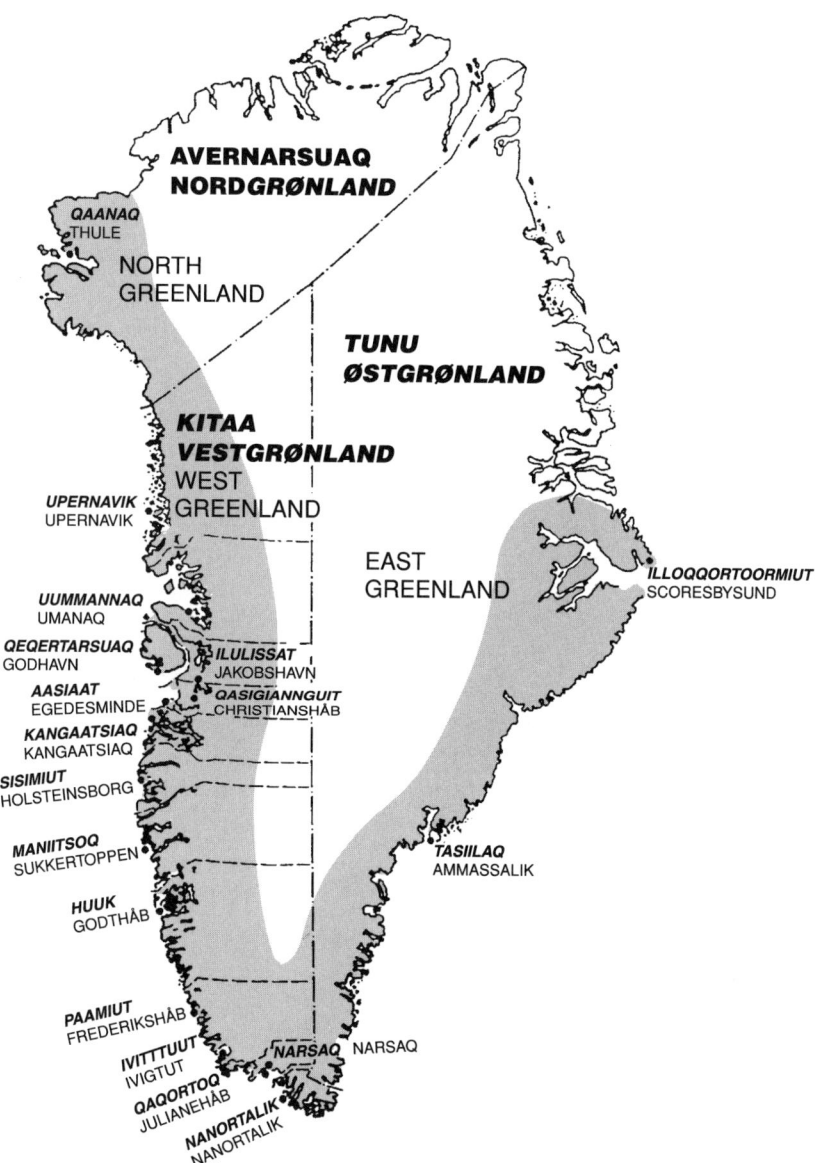

Community-based Whaling in Greenland.

Whaling and Sustainability in Greenland

Richard A. Caulfield
Department of Rural Development
University of Alaska, Fairbanks
1994

Introduction

In the aftermath of the U.N. Conference on Environment and Development (UNCED), the world's attention focuses increasingly on fostering economic development that is both sustainable and equitable. 'Sustainable development', according to the World Commission on Environment and Development (1987), is that which meets the 'needs of the present without compromising the ability of future generations to meet their own needs.' While the concept is somewhat problematic, it nevertheless focuses our attention on both maintaining the health of ecological systems and meeting the essential needs of human societies.

In advancing sustainability, ecologists today are seeking a better understanding of human practices that protect biodiversity and promote healthy marine and terrestrial ecosystems. The same is true of anthropologists, sociologists, and political scientists who seek to understand what types of human-environment interactions promote sustainability and equitability. The intent of these complementary processes in the scientific community is to provide clues about how human societies can modify their activities so that sustainable practices are promoted and unsustainable practices are discouraged.

Richard A. Caulfield

This paper draws upon these efforts to assess the role of whaling in a sustainable future for Greenlandic communities. Greenlanders take minke and fin whales under the IWC aboriginal subsistence whaling regime. This paper examines Greenlandic whaling using a framework developed by a group of social scientists (including the author) published in an article in *Oceans and Coastal Management* (Young *et al.* 1994). In the article, the authors examine 'historically-based practices of socially-defined human groups that value whaling activities on a multi-dimensional basis.' They do so to determine which whaling practices might be considered sustainable and equitable, and those which might not. Based upon extensive discussion, these social scientists conclude that some forms of whaling can be sustainable where users and managers recognize the importance of 'territoriality' and the maintenance of social institutions that effectively restrict user access to commonly valued, used, and managed resources.' They suggest five criteria or questions for identifying sustainable and equitable whaling practices:

- Is whaling conducted within *socially defined groups?*
- Is whaling conducted within *identified territorial limits?*
- Are whaling practices *socially reproducible over time?*
- Are whaling practices *valued multi-dimensionally?*
- Can management regimes *ensure biological sustainability?*

Importantly, the authors also recognize that, for whaling in indigenous societies, another question must be asked: Is whaling recognized as a cultural right under provisions of international law? In what follows, I analyze Greenlandic whaling in light of these criteria, drawing upon earlier reports (e.g., Caulfield 1991, 1993; Josefsen, this volume; Larsen and Hansen, this volume) and upon unpublished data.

Whaling Within Socially-defined Groups: Greenland's Inuit Society

According to Young *et al.* (1994), sustainable whaling 'must be carried out by groups who share a common culture or social bonds and whose maintenance is demonstrably dependent upon whale harvesting or the consumption of whale products.' That is, whaling must reinforce social and cultural patterns, and play a central role in maintaining social relations within the group.

For all of its 4,000 year history, Greenlandic Inuit society has been a hunting society. In remote Arctic settlements, whales, seals, and other renewable marine resources have provided the basis for a system of social organization based upon bilateral kinship ties within extended families. Sharing and exchange of wild foods and other local products are vital elements of this system. The social organization of Greenlandic society today continues to be kin-based. Nuclear and extended families continue to form the basis for social organization in hunting and other subsistence pursuits. Greenland is a society of 55,000 people, over 85% of whom identify themselves as Inuit and who speak Kalaallisut, the Greenlandic Inuit language. While Greenlandic settlements are widely dispersed along the country's extensive coastline, there is a strong and perhaps a growing sense of common identity, e.g., shared feeling of being Kalaalliit, or Greenlanders. The development of Greenland's system of Home Rule within the Danish realm has fostered a sense of collective solidarity as new technologies make communications and transportation more efficient and as Greenlanders become empowered to exercise greater control over their society.

Allocation of Greenland's minke and fin whale quotas reflects this kin-ordered system of social organization. Once the IWC makes its determination about Greenland's aboriginal subsistence whaling quotas, the Home Rule government allocates whale quotas to specific communities in consultation with Greenland's organization of municipal governments (KANUKOKA) and its national hunters' and fishers' organization (KNAPK). Only full-time hunters among Greenlanders are eligible to catch whales. This limits whaling to the most active hunters, those who have a close affiliation to Greenlandic society.

Furthermore, many of Greenland's 18 municipalities have their own criteria for allocating whales equitably when quotas are insufficient for local need. Most commonly, allocations are made to those who have few other income opportunities. Municipal authorities strive to ensure that those who do not receive approval to take a whale in a given year do receive approval in subsequent years. Likewise, municipalities ensure that allocations are shared equitably between larger towns and smaller communities. Thus, Greenland's catches of minke and fin whales (Table 1), is consistent with IWC quotas, and reflect both the need for sustainable harvests and equitable distribution of quotas throughout Greenlandic society.

Table 1.
Greenlandic Catch of Minke and Fin Whales, 1992 and 1993.

Municipality	Minke catch 1992	Fin whale catch 1992	Minke catch 1993	Fin whale catch 1993
Nanortalik	12	1	9	1
Qaqortoq	8	1	8	1
Narsaq	7	1	5	0
Paamiut	12	1	10	1
Nuuk	11	4	17	5
Maniitsoq	14	2	13	1
Sisimiut	10	0	13	0
Kangaatsiaq	3	1	5	0
Aasiaat	8	2	7	1
Qasigiannguit	2	1	4	0
Ilulissat	3	1	4	3
Qeqertarsuaq	3	1	4	0
Uummannaq	2	0	1	0
Upernavik	1	0	1	0
Total West Greenland	96	16	101	13
East Greenland	8	0	9	0

Whaling Within Territorial Limits: Coastal Catches for Local Consumption

Social scientists studying whaling communities note that sustainable and equitable whaling practices typically are confined to 'operations associated with specific shore-based communities' (Young et al. 1994). This observation reflects the importance of territoriality in promoting stewardship of marine resources and effective management regimes. In Greenland, minke and fin whaling is carried out from local communities ranging in size from about 50 to 13,000 people. The products of that whaling are strictly for Greenlandic consumption. By law, no whale products can be exported.

Minke whales taken in West Greenland are typically caught by fishing vessels up to 40 feet in length equipped with harpoon cannons. They are also caught in smaller numbers by hunters in skiffs in what is known as a collective hunt (Caulfield 1991). The Home Rule government grants approval for the collective hunt in taking minke whales where no fishing vessels are available. Fin whales are caught solely using the larger fishing vessels. These vessels are primarily used in commercial fishing and are only occasionally used for whaling. In both cases, the craft used in whaling are ill-suited for extensive voyages away from land.

Whaling activities are typically carried out within 10 to 20 km offshore. Hunters and fishermen out in their boats catching food for their families keep close watch for whales. In many cases, hunters prefer to catch and flense whales within sight of the local community where they can get assistance from family and community members. Often, community members monitor the hunters' efforts on walkie-talkie radios. Hunters customarily flense a whale within 24 hours to avoid spoilage. In all municipalities, hunters know traditional sites where conditions are best for the flensing process and where flensing equipment is commonly kept. When a whale is towed to one of these sites, community members often come down to assist in processing the whale. In return, they may receive a share of the meat or the *mattak*, the whale's skin and first layer of blubber.

Social Reproducibility: Greenlandic Whaling Through the Generations

Sustainable and equitable whaling involves practices that are socially reproducible over time. That is, 'the rules governing whaling and the knowledge to engage successfully in whaling must ordinarily be handed down from generation to generation within the same community' (Young *et al.* 1994). Importantly, this process of handing down knowledge 'serves to reinforce relationships based on kinship or other alliance-creating institutions.'

In Greenland, detailed traditional knowledge about the marine environment and about whales and other marine resources is handed down from generation to generation. Greenlanders are taught at an early age that whales and other creatures from the sea are gifts from *Sassuma Arnaa*, the 'mother of the sea.' According to Greenlandic legend, she gives whales, seals, and other animals to hunters who show these animals proper respect.

Hunters typically learn about whaling from experienced elders who know about feeding areas, whale behavior, and effective and safe whaling practices. In many communities, oral traditions relate the exploits of a long-deceased *piniartorsuaq*, or great hunter, who excelled at whaling or other subsistence pursuits. This body of knowledge is usually shared only with those speaking *Kalaallisut* and is therefore usually inaccessible to non-Greenlanders.

The social organization of whaling itself reflects the kin-ordered nature of Greenlandic society. Figure 1, derived from the author's earlier

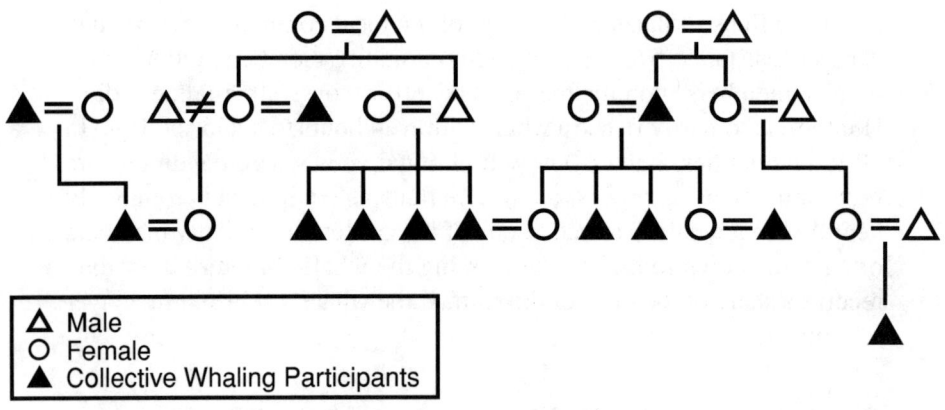

Figure 1. Simplified diagram showing kinship relations among participants in a collective whale hunt (Source: Caulfield 1991).

research in Disko Bay in West Greenland, illustrates the close kin relationships between participants in a collective hunt for a minke whale. As with indigenous environmental knowledge, these kin relationships are usually invisible to outsiders.

Valuing Greenlandic Whaling Multi-Dimensionally

A key difference between sustainable and unsustainable whaling practices may be the extent to which societies value whaling beyond mere economic considerations. That is, sustainable whaling is often embedded within the social fabric of a coastal community that has long-standing historical, socioeconomic, cultural, and nutritional relationships with marine resources. In these instances, whaling is not a means to maximize profits. In Greenland, hunters are not interested in developing whaling as a major industry. Rather, they are interested in whaling to meet nutritional needs, to sustain indigenous social and cultural systems and to generate sufficient cash to support local livelihoods. They seek, as Freeman (1993) points out, a means to 'sustain local social, cultural, and economic activity intergenerationally.'

In Greenland, whaling is embedded within sociocultural and economic systems extending back generations. Whaling has strong historical roots in Inuit society, roots which remain despite dramatic changes in whaling practices and technologies. Ecological dynamics, Euroamerican over-exploitation of whales and Danish colonial policies have all contributed to dramatic changes in Greenlandic whaling, particularly in the last 300 years. New whaling methods and technologies in use today, including the harpoon cannon and exploding penthrite harpoon, reflect these changes. However, perhaps the most striking thing about Greenlandic whaling is not how much it has changed, but that it has persisted despite these obstacles.

Greenlanders prize food from whales and other 'country foods', which they refer to as *kalaalimerngit,* or 'Greenlanders' food.' These foods constitute a substantial part of household diets and are integrally linked to Greenlandic identity. They are clearly distinguished from 'white man's food' or *qallunaamerngit,* which is considered inferior for those living in the Arctic. This distinction has important cultural significance. As Larsen and Hansen (this volume) point out, 'eating Greenlandic food is of great symbolic weight in determining whether a person is a true Greenlander.' Furthermore, the procurement, processing and sharing of *kalaalimerngit* reflects underlying systems of reciprocity and community solidarity that continue to be important in Greenlandic life today (Caulfield 1993, Petersen 1987).

Contemporary Greenlandic communities, like indigenous communities elsewhere in the Arctic, have mixed subsistence-cash economies.

In these communities, cash and subsistence sectors are mutually supportive. Cash generated through wage employment or other local activity underwrites subsistence production (Caulfield 1993). Distinguishing between 'commercial' and 'subsistence' sectors in Greenland's economy can be difficult at best. Indeed, Dahl (1989) argues that these sectors are so closely interrelated in Greenland that efforts to distinguish between them are meaningless.

This relationship between subsistence and cash is clearly revealed in the local production of country foods in Greenland, where hunters and fishermen catch a variety of marine resources, keeping some for household consumption, but also selling some for cash. These foods are sometimes sold on an individual basis within a community, but more often they are sold at a local *kalaaliaraq*, an outdoor kiosk located near the harbor in many Greenlandic towns. They may also be sold to Royal Greenland A/S, a Home Rule government-owned processing company. Greenlanders clearly prefer to buy fresh whale products and other country foods at the *kalaaliaraq* rather than from a local store. While store purchases may be necessary in winter months, the processing required makes the products less desirable.

Prices for foods sold at a *kalaaliaraq*, shown in Table 2, are negotiated annually between local authorities and the hunters' and fishers' association. This fixed-price system minimizes any tendency toward competition among hunters and ensures that community needs are met. Greenland's system for local food production helps reduce the country's dependence on the world economy. Even with the production of country foods, Greenland imports substantial quantities of protein in the form of meat products (Table 3). By supporting a local market for country foods, including whale products, the Home Rule government accomplishes multiple goals:

- it reduces imports, utilizes renewable rather than nonrenewable resources;
- enables Greenlanders to eat wholesome and nutritious foods, supports the economies of small settlements with few economic alternatives, and;
- promotes Greenlandic traditional hunting values.

Table 2.
Selected Prices (in U.S. $) for Locally Caught Products at Nuuk's *Kalaaliaraq (Outdoor Kiosk), 1991-1992.*

Product	Kalaaliaraq price ($/kg)
Seal meat (ringed seal)	4.62
Minke whale meat	7.69
Fin whale meat	6.15
Caribou meat	9.23
Eider duck	4.62

Source: Nuuk Fishermen's and Hunter's Association 1991

Biological Sustainability and Greenlandic Whaling

While the criteria discussed so far may be necessary for sustainable and equitable whaling, they alone are not sufficient. The literature on common property resources makes clear that they must be accompanied by biological research and monitoring systems that can take into account environmental changes affecting specific whale 'stocks.' Young *et al.* (1994) suggest that:

> the frequency and extent of such monitoring should be decided on a case-by-case basis. Given the inexact nature of fishery science, the best scientific judgments must necessarily allow for some (minimal) level of continuing disagreement among competent scientists regarding stock assessments.

Since the advent of Home Rule in 1979, Greenlanders have made significant improvements to management systems for monitoring and regulating whale harvests and for improving the efficiency of the kill. Home Rule officials have also made concerted efforts to educate hunters about whale population dynamics and to work cooperatively with KNAPK, the hunters' and fishers' organization. At the same time, marine mammal biologists in Greenland and Denmark, working closely with other scientists, have expanded considerably our knowledge of whale stocks in the North Atlantic.

Table 3.
Greenlandic Imports of Meat Products, 1991 and 1992 (in metric tons).

Type	1991	1992
Sausage (pølser)	575.1	558.8
Beef	682.0	647.6
Pork	1,548.4	1,507.0
Other	965.6	755.5
Total meat imports	3,771.1	3,468.9

(Source: Greenland Statistical Office 1994)

Laws governing Greenlandic whaling are enacted by the Greenlandic Parliament. Regulations implementing these laws are promulgated by the Home Rule administration in conjunction with local municipalities and KNAPK. Participants involved with whaling must have a full-time hunting license, reside permanently in Greenland, and have a close affiliation with Greenlandic society (Table 4). Furthermore, only hunters with a special license for whaling may take a minke or fin whale. This license obligates the hunter to follow Home Rule whaling regulations and to report the striking or killing of a whale to appropriate authorities.

Sanctions are imposed on hunters who violate these regulations. Home Rule authorities can fine individual hunters and confiscate any illegally taken whale products. Moreover, as an additional penalty, authorities can reduce quotas in subsequent years for the hunter's entire community. Because of the close kinship ties in Greenlandic communities, these actions can bring powerful social pressures to bear on hunters, encouraging them to avoid illegal hunting.

Greenland's management regime for whaling has developed considerably since the implementation of Home Rule in 1979. Perhaps one of the most significant developments is the cooperation hunters have shown in reducing the time it takes to kill a whale by using new technology. Since 1991, all hunters involved with fishing vessel whaling

Table 4.
Selected Greenlandic Whaling Regulations, 1993.

REGULATION	MINKE WHALING	FIN WHALING
Type of Hunt	• vessel whaling, OR • collective whaling (only with special dispensation)	• vessel whaling only
Hunter licensing	• full-time hunting license • permanent resident • close affiliation with Greenlandic society	• full-time hunting license • permanent resident • close affiliation with Greenlandic society

Whaling license	• required from municipality
	• dispensation for collective hunt when hunt has major significance for local community and where meat from vessel whaling is not available
		• required from municipality
		• can be issued to:
		  – one vessel ≥ 36 feet in length
		  – two vessels ≥ 30 feet in length

Season	• 1 April-31 December	• 1 January-31 December

Hunt Requirements	• no females with young may be taken
	• whale must be killed as quickly as possible
	• use of vessel with harpoon cannon (≥ 50mm)
	  – harpoon grenade required
	  – training for harpooner
	  – cannon in good condition
	  – registration and inspection of cannon by authorities
	  – equipped with winch
	OR
	• special dispensation for collective hunt
	  – minimum of 5 skiffs
		• no females with young may be taken
		• whale must be killed as quickly as possible
		• use of vessel with harpoon cannon (≥ 50mm)
		  – harpoon grenade required
		  – training for harpooner
		  – cannon in good condition
		  – registration and inspection of cannon by authorities
		  – equipped with winch
		• whale must be ≥ 15.2m

Table 4. (continued)
Selected Greenlandic Whaling Regulations, 1993.

– use 7.62mm rifles or larger – semi- or full automatic rifles prohibited – all skiffs equipped with hand harpoon, float, and ≥ 12mm line – designated hunt leader	
• all edible meat and *mattak* must be used	• all edible meat and *mattak* must be used
• catch data must be reported to authorities before sale of any whale products	• catch data must be reported to authorities before sale of whale products
• sample of whale meat and *mattak* must be provided for research.	• sample of whale meat and *mattak* must be provided for research.

(Sources: Greenland Home Rule Government 1992, 1993a, 1993b, 1993c.)

have been required to use the penthrite grenade, which kills whales more quickly than earlier methods. With support from the Home Rule government and KNAPK, hunters made the transition to this new technology with little objection, despite additional costs and the need for special training. These and other examples demonstrate that the Home Rule government's regime for managing whaling is increasingly effective and responsive to a wide range of concerns.

Whaling and Indigenous Rights

An additional issue not addressed directly by Young *et al.* (1994) is the relationship between whaling and indigenous rights. The authors do state that, in their judgment, aboriginal subsistence whaling practices currently recognized by the IWC are likely to be viewed as sustainable and equitable forms of whaling. Furthermore, they recognize that 'aboriginal whaling is protected in some states as a right that the relevant governments are obliged to uphold and protect' and that 'some principles of international law also support small-scale whaling when conducted by indigenous peoples.' These statements are consistent with the views of most Greenlanders, who believe that whaling is protected by provisions of international law, including Article 27 of the U.N. International Covenant on Civil and Political Rights, which uphold the right of minority peoples 'to enjoy their own culture.' As Greenland's premier, Lars Emil Johansen (1991:7) states,

> For us who live in Greenland, hunting and trapping are natural elements in our life, from cradle to grave. The people of Greenland have always been hunters... It is a way of life which has permeated our myths and legends for thousands of years, and has formed our attitudes toward hunting and the relationship with the natural environment... Our lives as hunters and trappers are at the very root of our identity. Our attitude toward nature and its resources is quite distinct from the attitudes of people in the industrialized world.

Thinking Globally, Acting Locally: Sustainable Whaling in Greenland

When it comes to whaling and the use of other marine mammals, Greenlanders abide by the principle of 'thinking globally and acting locally.' While Greenland's ties to the world economy are increasing, most Greenlanders continue to identify themselves first and foremost as hunters, fishers and trappers. At a time when many people in industrialized countries are beginning to appreciate the importance of being rooted in a place and of being stewards of local resources, Greenlanders draw upon 4,000 years of history as coastal people, reliant on renewable resources from the sea as the basis for their survival.

As Greenland's population grows and as Greenlanders continue to become accustomed to a higher standard of living, they will undoubtedly seek new opportunities to create jobs and to expand ties with other countries. But even with these changes, many Greenlanders, particularly those living in smaller outlying settlements, will continue to rely on whales and other marine resources for nutritional, sociocultural and economic benefits. Greenlanders today realize that survival in the modern Arctic requires 'thinking globally' about environmental pollution in the Arctic and about global warming, but it also requires 'acting locally,' being careful stewards and respectful users of renewable resources. They have demonstrated clearly their willingness to cooperate with the IWC in matters of resource conservation. They have, for example, accepted a zero quota on the take of humpback whales because of international concern about 'stock' levels. They have also worked diligently to improve whaling practices in response to concerns about humane killing. And, Home Rule officials continue to work closely with KNAPK and hunters to improve hunting practices and to gain increased compliance with whaling regulations.

In Greenland today, there need not be a contradiction between protecting the health of whale stocks and meeting the essential needs of human beings. Whaling can be a major contributor to sustainable development in Greenland. However, to achieve this goal there must be greater understanding of the characteristics of mixed subsistence-cash economies. There must also be an effective management regime to monitor whale number interactions and to ensure compliance with appropriate regulations. And, perhaps most importantly, there must be recognition of the fact that human societies relate to their environments in different ways. This diversity offers strength for all humankind, for out of it may come solutions to some of our most pressing problems.

References

Caulfield, R.A. 1991. *Qeqertarsuarmi arfanniarneq: Greenlandic Inuit whaling in Qeqertarsuaq Kommune, West Greenland.* Report presented to the International Whaling Commission Technical Committee. TC/43/AS4.

Caulfield, R.A. 1993. Aboriginal subsistence whaling in Greenland: The case of Qeqertarsuaq Municipality in west Greenland. *Arctic* 46(2): 144-155.

Dahl, J. 1989. The integrative and cultural role of hunting and subsistence in Greenland. *Études/Inuit/Studies* 13(1): 23-42.

Freeman, M.M.R. 1993. Introduction. Community-based whaling issue. *Arctic* 46(2): iii-iv.

Greenland Home Rule Government. 1992. *Hjemmestyrets bekendtgørelse nr. 42 af 18. december 1992 om rapportering ved fangst og anskydning af hvaler.* Grønlands Hjemmestyre, Nuuk.

Greenland Home Rule Government. 1993a. *Hjemmestyrets bekendtgørelse nr. 18 af 22. juli 1993 om tilladelse til erhvervsmæssig fangst og jagt (erhvervsjægere).* Grønlands Hjemmestyre, Nuuk.

Greenland Home Rule Government. 1993b. *Hjemmestyrets bekendtgørelse nr 20 af 29. juli 1993 om fangst af store hvaler.* Grønlands Hjemmestyre, Nuuk.

Greenland Home Rule Government. 1993c. *Hjemmestyrets bekendtgørelse nr. 19 af 22. juli 1993 om ikke-erhvervsmæssig fangst og jagt (fritidsjægere).* Grønlands Hjemmestyre, Nuuk.

Greenland Home Rule Government. 1994. *1992 aamma 1993-imi Nunatsinni arfattat.* Unpublished report, Nuuk.

Greenland Statistical Office (Gronland Statistiske Kontor). 1994. *Indførsel af kød 1991-92 i tons.* Unpublished, Nuuk.

Johansen, L.E. 1991. Nature: Our survival. In *Nature conservation in Greenland.* H.J. Helms (editor), Atualliorfik, Nuuk.

Josefesen, E. 1990. *Cutter hunting of minke whales in Qaqortoq (Greenland).* (this volume).

Larsen, S.E. and K.G. Hansen. 1990. *Inuit and whales at Sarfaq (Greenland).* (this volume).

Nuuk Fishermen's and Hunter's Association (Nungme Aulisartut Piniartdlo Pekatigit). 1991. *Tuniniaarsarfimmi akit 1991-mi.* Unpublished, Nuuk.

Petersen, R. 1987. *Communal aspects of preparing for whaling, the hunt itself, and of the ensuing products.* (this volume).

World Commission on Environment and Development. 1987. *Our Common Future.* Oxford University Press, Oxford and New York.

Young, O.R., M.M.R. Freeman, G. Osherenko, R.R. Anderson, R.A. Caulfield, R.L. Freidheim, S.J. Langdon, M. Ris, and P.J. Usher. 1994. Subsistence, Sustainability, and Sea Mammals: Reconstructing the International Whaling Regime. *Ocean & Coastal Management* 23:117-127.